CC
GE

Facts

HarperCollins*Publishers*

HarperCollins Publishers
PO Box, Glasgow G4 0NB, Scotland
www.collins-gem.com

First published 1983
Fourth edition 1999

Reprint 10 9 8 7 6 5 4 3 2 1

ISBN 0 00 472355-4

Printed and bound in Great Britain by
Caledonian International Book Manufacturing Ltd,
Glasgow

Introduction

Collins Gem *Basic Facts* is a series of illustrated GEM dictionaries in important school subjects. This new edition has been revised and updated to widen the coverage of the subject and to reflect recent changes in the way it is taught in the classroom.

Bold words in an entry identify key terms which are explained in greater detail in entries of their own; important terms which do not have separate entries are shown in *italic* and are explained in the entry in which they occur.

Other titles in the series include:

Gem *Mathematics Basic Facts*
Gem *Chemistry Basic Facts*
Gem *Physics Basic Facts*
Gem *Science Basic Facts*
Gem *Computers Basic Facts*
Gem *Twentieth-Century History Basic Facts*
Gem *Biology Basic Facts*
Gem *Business Studies Basic Facts*

abrasion   The wearing away of the landscape by rivers, **glaciers**, the sea or wind, caused by the **load** of debris that they carry. See also **corrasion**.

abrasion platform   See **wave-cut platform**.

accessibility   A measure of the ease and efficiency with which a location can be reached. Central locations are highly accessible; peripheral ones are not.

acid rain   Rain that contains a high concentration of pollutants, notably sulphur and nitrogen oxides. These pollutants are produced from factories, power stations burning **fossil fuels** and car exhausts. Once in the **atmosphere**, the sulphur and nitrogen oxides combine with moisture to give sulphuric and nitric acids which fall as corrosive rain. The results of acid rain have been devastating to many lakes, forests and buildings in Scandinavia and Germany, due to **prevailing winds** in this part of the northern hemisphere blowing from a southwesterly direction, bringing acid rain formed in the industrial centres of western Europe. See **pH**.

administrative region   A defined area in which governmental, commercial and other organizations carry out administrative functions. Examples include the regions of local health authorities and water companies, and commercial sales regions.

adult literacy rate   A percentage measure showing the proportion of an adult population that can read. It is one of the measures used to assess the level of development of a country.

aerial photograph   A photograph taken from above the ground. There are two types of aerial photograph – a vertical photograph (or 'bird's eye view') and an oblique photograph where the camera is held at an angle. Aerial photographs are often taken from aircraft and provide useful information for map making and surveys. Compare **satellite image**.

afforestation   The conversion of open land to forest; especially, in Britain, the planting of coniferous trees in upland areas for commercial gain. Coniferous afforestation is currently a controversial issue in the Flow Country of northern Scotland, where it has destroyed large parts of this unique blanket bog area which is the **habitat** of many rare plants, birds and animals. **Conservation** efforts are now being made to halt the rate of afforestation and to protect what is left of this **ecosystem**.

   In areas of the world at risk from **desertification**, the planting of suitable trees helps to slow down

soil erosion, and thus the expansion of the desert, by both binding the soil and providing **rainfall interception**. The trees also shelter the land, and any crops, from the force of the wind, and produce humus to increase the fertility of the soil (see **horizon**). Compare **deforestation**.

age and sex structure   The classification of the elements of a national or regional population according to sex and age groups. Age and sex structure determines the shape of a **population pyramid**.

agglomerate   A mass of coarse rock fragments or blocks of lava produced during a volcanic eruption.

agglomeration   The tendency for firms with related products to locate close to each other in order to reduce transport costs and other overheads: for example, the motor manufacture and component industries, and the oil refining and petrochemical industries.

agribusiness   Modern **intensive farming** which uses machinery and **artificial fertilizers** to increase **yield** and output. Thus agriculture resembles an industrial process in which the general running and managing of the farm could parallel that of large-scale industry. Many farms in East Anglia could be called agribusinesses. There small family farms have been bought up by large food-processing companies located near the farms, where the farm produce is canned, processed, frozen, etc.

A 'branch' of agribusiness is factory farming, i.e. the intensive rearing of animals such as pigs, cattle and poultry. Many of these animals spend their lives penned in special rearing units where food, water, light and temperature are carefully controlled. Thus factory farming can be seen as a production line, the sole aim being to increase output and profit. Many people disagree with this form of farming, but it is argued that less intensive methods mean higher prices for the consumer.

agriculture Human management of the **environment** to produce food. The numerous forms of agriculture fall into three groups: **commercial agriculture**, **subsistence agriculture** and **peasant agriculture**. See also **agribusiness**.

aid The provision of finance, personnel and equipment for furthering economic development and improving standards of living in the **Third World**. Most aid is organized by international institutions (e.g. the United Nations), by charities (e.g. Oxfam) (see **non-governmental**

organizations (NGOs); or by national governments. Aid to a country from the international institutions is called *multilateral aid*. Aid from one country to another is called *bilateral aid*.

Aid can be *short-term* to meet an immediate crisis, *long-term* to further development, or *tied*, linking aid expenditure directly back to the donating country.

alluvial fan A cone of **sediment** deposited at an abrupt change of slope; for example, where a post-glacial stream meets the flat floor of a **U-shaped valley**. Alluvial fans are also common in arid regions where streams flowing off **escarpments** may periodically carry large **loads** of sediment during **flash floods**

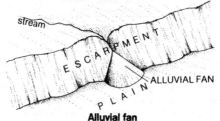

Alluvial fan

alluvium Material deposited by a river in its middle and lower course. Alluvium comprises

silt, sand and coarser debris eroded from the river's upper course and transported downstream. Alluvium is deposited in a graded sequence: coarsest first (heaviest) and finest last (lightest). Regular floods in the lower course create extensive layers of alluvium which build up to considerable depth on the **flood plain**. Some of the world's most fertile lands are found on alluvial flood plains.

alp A gentle slope above the steep sides of a glaciated valley, often used for summer grazing. See also **transhumance**.

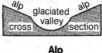

Alp

amenity resources Those **resources** which provide an opportunity for recreation and leisure pursuits. The **national parks**, forests and the coastline are good examples of these, as are areas of parkland and open space in cities.

anemometer An instrument for measuring the velocity of the wind. An anemometer should be fixed on a post at least 5 m above ground level.

The wind blows the cups round and the speed is read off the dial in km/hr (or knots).

dial

Anemometer A cup anemometer.

anthracite A hard form of **coal** with a high carbon content and few impurities.

anticline An arch in folded **strata**; the opposite of **syncline**. See **fold**.

anticyclone An area of high atmospheric pressure with light winds, clear skies and settled **weather**. In summer anticyclones are associated with warm and sunny conditions; in winter they bring frost and fog as well as sunshine.

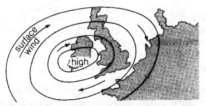

surface wind

high

Anticyclone Weather map showing an anticyclone over the British Isles.

appropriate technology Techniques and equipment which are appropriate to the immediate needs of a **developing country**. For example, ox-drawn planters and reapers, made of local materials, may be more useful in developing **agriculture** in such countries than tractors and combines, which are expensive to run, difficult to maintain in remote regions, and may cause unemployment. Appropriate technology stresses the need for low cost, straightforward and **labour-intensive** projects, as opposed to the **capital-intensive, high-technology approach**. See also **intermediate technology**.

aquifer See **artesian basin**.

arable farming The production of cereal and root crops – as opposed to the keeping of livestock.

arête A knife-edged ridge separating two **corries** in a glaciated upland. The arête is formed by the progressive enlargement of corries by **weathering** and **erosion**. See fig. on next page. See also **pyramidal peak**.

artesian basin This consists of a shallow **syncline** with a layer of **permeable rock**, e.g. chalk, sandwiched between two layers of **impermeable rock**, e.g. clay. Where the permeable

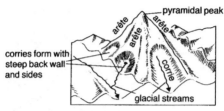

Arête

rock is exposed at the surface, rainwater will enter the rock and the rock will become saturated. This is known as an *aquifer*. Boreholes can be sunk into the structure to tap the water in the aquifer. If there is sufficient pressure of water within the aquifer, the water will rise freely to the surface. The London Basin consists of a shallow syncline formed of a layer of chalk between two layers of clay. Because London's **water table** has dropped considerably in recent years (due to demand by the water companies and industry) water now has to be pumped to the surface, as the pressure of water in the aquifer is too low for it to flow out freely. See fig. overleaf.

artificial fertilizer A chemical product containing one or all of the following: nitrogen, potash, phosphates and trace elements, which are derived from petrochemicals, natural phosphate

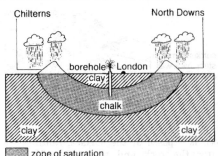

zone of saturation

Artesian basin The London Basin.

deposits and other industrial sources. In contrast **organic fertilizers** include animal dung, rotted vegetation (compost) and animal derivatives such as bone meal. Artificial fertilizers, if leached into streams and rivers, can cause problems of **eutrophication**.

artificial fibres   Textile materials, e.g. nylon, rayon, viscose, and polyester, made from hydrocarbons, derived from such materials as oil, **coal** and wood fibres, in contrast with natural fibres such as wool and cotton.

assembly industry   A firm which assembles components into a finished product. For example, the motor-car industry assembles parts such as engines, bodies, wheels, windscreens and electrical components into a final product. The individual components are made by a number of other firms. Assembly industry is distinguished from a manufacturing industry such as steel-making, which uses **primary** products such as **coal, limestone** and iron ore.

asymmetrical fold Folded **strata** where the two **limbs** are at different angles to the horizontal.

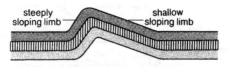

steeply sloping limb — shallow sloping limb

Asymmetrical fold Structure.

atmosphere The air which surrounds the Earth, which is up to 15 km in thickness at the equator and less thick at higher latitudes. The atmosphere comprises oxygen (21%), nitrogen (78%), carbon dioxide, argon, helium and other gases in minute quantities.

attrition The process by which a river's **load** is eroded through particles, such as pebbles and boulders, striking each other.

backwash   The return movement of seawater off the beach after a wave has broken. See also **swash** and **longshore drift**.

balance of trade   See **international trade**.

barchan   A type of crescent-shaped sand dune formed in desert regions where the wind direction is very constant. Wind blowing round the edges of the dune causes the crescent shape, while the dune may advance in a downwind direction as particles are blown over the crest.

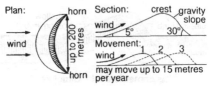

Barchan Formation.

bar graph   A graph on which the values of a certain variable are shown by the length of

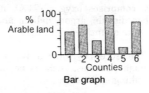

Bar graph

shaded columns, which are numbered in sequence. Compare **histogram**.

barograph A aneroid **barometer** connected to an arm and inked pen which records pressure changes continuously on a rotating drum. The drum usually takes a week to make one rotation.

barometer An instrument for measuring atmospheric pressure. There are two types, the *mercury barometer* and the *aneroid barometer*. The mercury barometer consists of a glass tube containing mercury which fluctuates in height as pressure varies. The aneroid barometer is a small metal box from which some of the air has been removed. This box expands and contracts as the air pressure changes. A series of levers joined to a pointer shows pressure on a dial.

barrage A type of dam built across a wide stretch of water, e.g. an estuary, for the purposes of water management. Such a dam may be intended to provide water supply, to harness wave energy or to control flooding, etc. There is a large barrage across Cardiff Bay in South Wales.

barrier beach A long narrow beach that extends across a bay and which can lead to the formation of a **lagoon** on the landward side.

basalt A dark, fine-grained extrusive **igneous rock** formed when **magma** emerges onto the Earth's surface and cools rapidly. The Giant's Causeway, Northern Ireland, is composed of basalt. As basalt cools a hexagonal pattern of jointing may occur as it contracts – hence the hexagonal basalt columns of the Giant's Causeway and of Fingal's Cave, island of Staffa, Scotland. A succession of basalt **lava flows** may lead to the formation of a **lava plateau**.

base flow The water flowing in a stream which is fed only by **groundwater**. During dry periods it is only the base flow which passes through the stream channel.

basin of internal drainage In desert regions depressions (sometimes below sea level) may occur, from which there is no natural outlet. Intermittent streams drain to the centre of the basin and **drainage** occurs by evaporation. Lakes occupying such basins fluctuate in depth and area according to the balance of inflow and

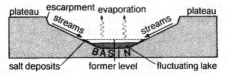

Basin of internal drainage

evaporation. Extensive salt deposits (e.g. halite, gypsum) may be left behind as lakes evaporate, e.g. Dead Sea, Makgadikgadi Pan (Botswana).

batholith A large body of igneous material intruded into the Earth's **crust**. As the batholith slowly cools, large-grained **rocks** such as **granite** are formed. Batholiths may eventually be exposed at the Earth's surface by the removal of overlying rocks through **weathering** and **erosion**. Dartmoor and Exmoor in Southwest England are offshoots (bosses) from a batholith.

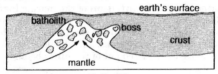

Batholith

bay An indentation in the coastline with a **headland** on either side. Its formation is due to the more rapid **erosion** of softer rocks. (See fig. overleaf.)

bay bar A bank of sand or shingle extending as a barrier almost or totally across a **bay**, caused by **longshore drift**.

beach A strip of land sloping gently towards

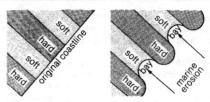

Bay How a bay is formed.

the sea, usually recognized as the area lying between high and low tide marks.

Sand is the final product of marine **erosion**, formed largely by **attrition** of the marine **load**. Beach deposits may be varied, comprising sand and shingle deposited by the sea, banks of pebbles deposited in storm tides, and boulders which have been weathered out of the **cliffs** behind.

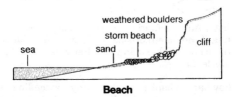

Beach

bearing   A compass reading between 0 and 360 degrees, indicating direction of one location from another.

Bearing   The bearing of B from A is 110°.

Beaufort wind scale   An international scale of wind velocities, ranging from 0 (calm) to 12 (hurricane). See **Appendix 1**.

bedding plane   The division between two **strata** of rock, generally indicating the boundary between earlier and later periods of **deposition**.

bergschrund   A large **crevasse** located at the rear of a **corrie** icefield in a glaciated region, formed by the weight of the ice in the corrie dragging away from the rear wall as the **glacier** moves downslope.

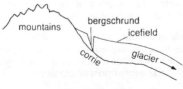

Bergschrund

biodiversity   The existence of a wide variety of plant and animal species in their natural environment. Conservationists are concerned that pollution and other human interventions such as intensive agriculture and genetically modified crops are reducing biodiversity and hence the stability of **ecosystems**.

biogas   The production of methane and carbon dioxide, which can be obtained from plant or crop waste. Biogas is an example of a renewable source of energy (see **renewable resources, nonrenewable resources**).

biomass   The total number of living organisms, both plant and animal, in a given area.

biosphere   The part of the Earth which contains living organisms. The biosphere contains a variety of **habitats**, from the highest mountains to the deepest oceans.

birth rate   The number of live births per 1000 people in a population per year.

bituminous coal   Sometimes called housecoal – a medium-quality **coal** with some impurities; the typical domestic coal. It is also the major fuel source for **thermal power stations**.

block faulting   The dissection of a region by an

Block faulting A series of block mountains and rift valleys.

extensive system of vertical or semi-vertical **faults**. The faults divide the landscape into a series of blocks which may be relatively uplifted or depressed, to produce a series of **block mountains** (*horsts*) and **rift valleys**.

block mountain or **horst** A section of the Earth's **crust** uplifted by faulting. Mt. Ruwenzori in the East African Rift System is an example of a block mountain.

blowhole A crevice, **joint** or **fault** in coastal rocks, enlarged by marine **erosion**. A blowhole often leads from the rear of a cave (formed by wave action at the foot of a **cliff**) up to the cliff top. As waves break in the cave they erode the roof at the point of weakness and eventually a hole is formed. Air and sometimes spray are forced up the blowhole to erupt at the surface. At times wave action can be so strong that bursts of seawater are sent through the blowhole, enlarging it further. See diagram overleaf.

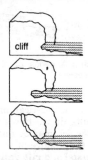

Blowhole Formation.

blue-collar worker A worker who is either a manual worker or who works in potentially 'dirty conditions'. The term 'blue-collar' derives from the wearing of dark-coloured overalls which show less dirt than light-coloured clothing. Compare **white-collar worker**.

blue ice Heavily compressed ice at the bottom of a **corrie** icefield or **glacier**, the expulsion of air caused by the compression making the ice appear blue. Contrast this with **white ice**.

bluff See **river cliff**.

boulder clay or **till** The unsorted mass of debris

dragged along by a **glacier** as *ground moraine* and dumped as the glacier melts. Boulder clay may be several metres thick and may comprise any combination of finely ground 'rock flour', sand, pebbles or boulders.

bourne A spring with a fluctuating origin on the **dip slope** of a **chalk escarpment**. In wet seasons, the spring may emerge higher up the dip slope, i.e. when the **water table** is higher. Drier weather will cause a drop in the level of the water table and the spring will then emerge further down the dip slope.

The term 'bourne' is commonly applied to streams in the chalk regions of southern England.

Bourne

BP *abbrev. for* Before Present, a term used in geological time scales to denote any time before the present.

break-of-bulk-point A point in freight carriage where goods are off-loaded from one mode of transport and loaded onto another.

A port is a major break-of-bulk-point, as is a railhead. Minimizing break-of-bulk-points is one way of reducing costs in transport systems.

breakwater or **groyne**   A wall built at right angles to a beach in order to prevent sand loss due to **longshore drift**.

breccia   Rock fragments cemented together by a matrix of finer material; the fragments are angular and unsorted. An example of this is volcanic breccia, which is made up of coarse angular fragments of **lava** and **crust** rocks welded by finer material such as ash and **tuff**.

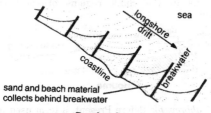

sand and beach material
collects behind breakwater

Breakwater

brownfield site   A site for development in an urban area that has previously been used (for example, the site of a demolished factory). Compare **greenfield site**.

Burgess model or **concentric theory**   A model
of urban structure formulated by E. W. Burgess
in 1923 and based on Chicago, in which five
major zones of urban land use are proposed. This
model, together with the **sector model** and
multiple nuclei model, can provide a basis for
looking at city structure in the developed world.
Compare **shanty town**.

1: central business
   district

2: transition zone
   (factory zone)

3: workingmen's
   homes

4: residential zone

5: commuter zone

Burgess model

bush fallowing or **shifting cultivation**   A
system of **agriculture** in which there are no
permanent fields. For example in the **tropical
rain forest**, remote societies cultivate forest
clearings for one year and then move on.

   The system functions successfully when forest
regeneration occurs over a sufficiently long
period to allow the soil to regain its fertility. See
diagram overleaf.

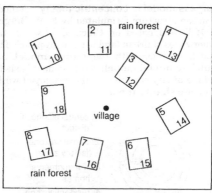

cultivated in year 1

and in year 10

Bush fallowing A typical bush fallowing regime.

business park An out-of-town site accommodating offices, high-technology companies and light industry. The estate is usually built on a *greenfield* site (one not previously built on), with landscaped grounds and low-density buildings.

It is typical for 70% of the site to be open, with grass, flowerbeds and trees, to create a pleasant environment for the workers. Unlike a **science park**, a business park does not have a direct link with a university. Some business parks also contain retail outlets such as hypermarkets and DIY stores, and some even include leisure complexes.

butte An outlier of a **mesa** in arid regions.

caldera A large crater formed by the collapse

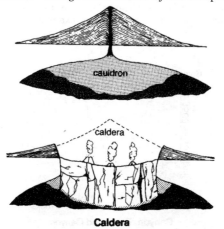

Caldera

of the summit cone of a **volcano** during an eruption. The caldera may contain **subsidiary cones** built up by subsequent eruptions, or a crater lake if the volcano is extinct or dormant.

canyon A deep and steep-sided river valley occurring where rapid vertical **corrasion** takes place in arid regions. In such an **environment** the rate of **weathering** of the valley sides is slow. If the **rocks** of the region are relatively soft then the canyon profile becomes even more pronounced. The Grand Canyon of the Colorado River in the USA is the classic example.

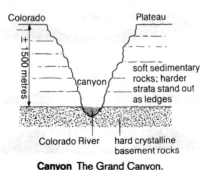

Canyon The Grand Canyon.

capital intensive Relating to an operation in which high productivity is achieved through high investment; for example, **market gardening** is a capital intensive form of **agriculture** identified by high investment in equipment such as greenhouses and heating and irrigation systems. See also **labour intensive**.

cap rock 1. (also called **fall maker**) A stratum of resistant **rock** at the lip of a **waterfall**.
  2. Hard rock protecting the top of a **mesa**.

catchment 1. In **physical geography**, an alternative term to **river basin**.
  2. In **human geography**, an area around a town or city – hence 'labour catchment' means the area from which an urban workforce is drawn.

cavern In **limestone** country, a large underground cave formed by the dissolving of limestone by subterranean streams. See also **stalactite, stalagmite**.

CBD (Central Business District) This is the central zone of a town or city, and is characterized by high **accessibility**, high land values and limited space. The visible result of these factors is a concentration of high-rise buildings at the city centre. The CBD is dominated by retail and business **functions**, both of which require maximum accessibility.

census The collection of information, used in particular to establish the accurate population of a country. In the UK, a questionnaire is sent to all **households** every 10 years by the Office of Population Censuses and Surveys.

   Information required by the census includes: the age and sex of the occupants; which people in the household work; where they work and how they travel to work; the number of rooms in the house; the nature of the **household amenities**; and whether the house has a garage. This information is used to assess the future needs of particular areas and households, e.g. whether new schools or old people's homes are likely to be required in the future. It is illegal not to fill in the census form.

central place A **settlement** offering goods and **services** to a surrounding population.

central place theory Walter Christaller's 1933 model of **settlement** location and distribution. Christaller envisaged a **settlement hierarchy** in which small **central places** would offer a limited range of everyday goods and **services** to a small surrounding population, and large central places would offer a large number of goods and services, many of them of a specialized nature, to a large surrounding population. Thus the **sphere of influence** of the large central places would be considerably greater in extent than that of the small central places. Christaller proposed several orders of central places. The diagram shows a typical Christaller hierarchy for an idealized and uniform **environment**.

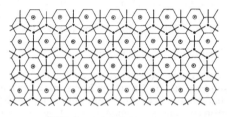

- • first order central place

⬡ first order sphere of influence

⊙ second order central place

◇ second order sphere of influence

further orders can be added

Central place theory

CFCs (Chlorofluorocarbons) Chemicals used in the manufacture of some aerosols, the cooling systems of refrigerators and fast-food cartons.

These chemicals are harmful to the **ozone** layer and their use is being greatly reduced.

chalk   A soft, whitish **sedimentary rock** formed by the accumulation of small fragments of skeletal matter from marine organisms; the rock may be almost pure calcium carbonate as a result. In Britain, chalk occurs in the low hills of the south and east, for example in the Downs and the Chilterns. Due to the **permeable** and soluble nature of the rock, there is little surface **drainage** in chalk landscapes.

chernozem   A deep, rich soil of the plains of southern Russia. The upper **horizons** are rich in lime and other plant nutrients; in the dry **climate** the predominant movement of **soil** moisture is upwards (contrast with *leaching*), and lime and other chemical nutrients therefore accumulate in the upper part of the **soil profile**.

cirrus   High, wispy or strand-like, thin **cloud** associated with the advance of a **depression**.

clay   A soil composed of very small particles of **sediment**, less than 0.002 mm in diameter. Due to the dense packing of these minute particles, clay is almost totally impermeable, i.e. it does not allow water to drain through. Clay soils very rapidly waterlog in wet weather.

cliff A steep rockface between land and sea, the profile of which is determined largely by the nature of the coastal rocks. For example, resistant rocks such as **granite** (e.g. at Land's End) will produce steep and rugged cliffs. The dip of the **strata** influences the gradient of the cliff.

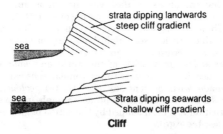

Cliff

climate The average atmospheric conditions prevailing in a region, as distinct from its **weather.** A statement of climate is concerned with long-term trends. Thus the climate of, for example, the Amazon Basin is described as hot and wet all the year round; that of the Mediterranean Region as having hot dry summers and mild wet winters. See **extreme climate, maritime climate**.

climatic change Fluctuations in the patterns of **climate** over long periods of time. There is indisputable evidence for changes in the world's climate: (a) *fossil records*: fossils of species such as the mammoth have been found in what are now temperate regions; (b) *topographical evidence*: in North Africa, now largely desert, it can be seen where rivers once ran, indicating that the area's rainfall was once much higher; (c) *historical records*: the freezing over of the River Thames was widely reported in the 17th and 18th centuries (see **Little Ice Age**); (d) *meteorological records*: records of the weather which exist for many areas of the world and some of which date from early times; comparisons can be made with present-day climate; (e) *geological evidence*: the landscape of, for example, much of Britain has been shaped by the action of glaciers (see **Ice Age**).

clint A block of **limestone**, especially when part of a **limestone pavement** where the surface is composed of clints and **grykes**.

cloud A mass of small water drops or ice crystals formed by the **condensation** of water vapour in the **atmosphere**, usually at a considerable height above the Earth's surface. There are three main types of cloud: **cumulus, stratus** and **cirrus**, each of which has many variations.

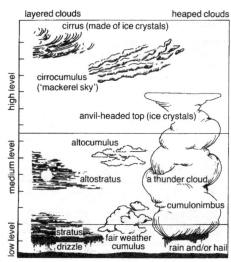

layered clouds   heaped clouds

high level

cirrus (made of ice crystals)

cirrocumulus ('mackerel sky')

anvil-headed top (ice crystals)

medium level

altocumulus

altostratus

a thunder cloud

cumulonimbus

low level

stratus

drizzle

fair weather cumulus

rain and/or hail

Cloud Different kinds of cloud.

coal A **sedimentary rock** composed of decayed and compressed vegetative matter. Coal in Britain dates from the Carboniferous period (about 350 million years ago) when tropical forest covered large areas of the land.

Coal is usually classified according to a scale of hardness and purity ranging from **anthracite** (the hardest), through **bituminous coal** and **lignite** to **peat**.

coastal marsh   A marsh formed by the growth of a **spit** across a **bay**, and the gradual silting up of the resulting **lagoon**. A well known example is Romney Marsh in southeast England. See diagram below.

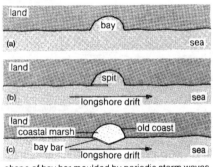

shape of bay bar moulded by periodic storm waves

The coastal marsh is consolidated in the following sequence:
a) **Siltation** – formation of mudflats covered at high tide
b) **Salt marsh** – colonization by reeds and other plants
c) **Freshwater marsh** – plants and soil build up above high water.

cold front See **depression**.

colonial influence The consequences of, for example, British colonial activity in many parts of the developing world. There are two broad areas where colonialism has had long-term and continuing influence: in government and administration, and in economic systems. Many former colonies have governments and bureaucracies modelled on institutions of the former colonial power, and this may not always be to the best advantage of the nation concerned. Economically, the majority of former colonies continue to be suppliers of **primary products** to the industrial world, a situation which may well perpetuate underdevelopment.

commercial agriculture A system of **agriculture** in which food and materials are produced specifically for sale in the market, in contrast to **subsistence agriculture**. Commercial agriculture tends to be **capital intensive**. See also **agribusiness**.

Common Agricultural Policy (CAP) The policy of the European Union to support and subsidize certain crops and methods of animal husbandry. It has been criticized for supporting inefficient farming techniques and also for encouraging overproduction. Minor reforms are under consideration.

common land   Land which is not in the ownership of an individual or institution, but which is historically available to any member of the community. Hence common grazing rights still remain in some upland areas of Britain. The English village green, where it survives, is often common land.

communications   The contacts and linkages in an **environment**. For example, roads and railways are communications, as are telephone systems, newspapers, and radio and television. See also **route**.

commuter zone   An area on or near to the outskirts of an urban area. Commuters are the most affluent and mobile members of the urban community and can afford the greatest physical separation of home and work. The commuter zone is thus the outer ring of suburbs (see **Burgess model**) and the villages beyond, the latter increasingly peopled by managers, professionals, etc., who travel daily to the city centre. See **dormitory settlement**.

comprehensive redevelopment   A policy of clearing substandard housing and completely rebuilding the area to create a new environment.

Comprehensive redevelopment usually takes place in **inner city** areas where conditions of overcrowding may be severe, and where houses are often mingled with industry and business. Comprehensive redevelopment aims to provide houses with good amenities, with areas of open space and parks, and which are away from industry and business. Some schemes have been successful but many schemes in the 1960s, where high-rise blocks of flats were included, have come in for much criticism.

concentric theory   See **Burgess model**.

concordant coastline   A coastline that is parallel to mountain ranges immediately inland. A rise in sea level or a sinking of the land will cause the valleys to be occupied by the sea and

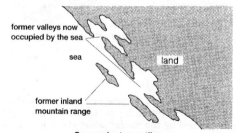

Concordant coastline

the mountains to become a line of islands. This has occurred along the coast of Croatia in the Adriatic sea. See fig. on previous page. Compare **discordant coastline**.

concealed coalfield   A coalfield in which **coal** measures are located at some depth beneath the overlying **strata**. In contrast is the *exposed coalfield* where coal measures outcrop at or near the surface, e.g. the Yorkshire coalfield.

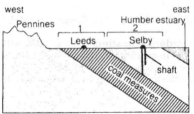

Concealed coalfield (1) Exposed coalfield — open cast or shaft mining since the late 18th century. (2) Concealed coalfield — recent development of deep mining.

condensation   The process by which cooling vapour turns into a liquid. **Clouds** for example are formed by the condensation of water vapour in the **atmosphere**.

coniferous forest   A forest of **evergreen** trees such as pine, spruce and fir. Natural coniferous forests occur considerably further north than forests of broad-leaved **deciduous** species as coniferous trees are able to withstand harsher climatic conditions. The **taiga** areas of the northern hemisphere consist of coniferous forests. Not all coniferous forests are indigenous to an area as some have been planted for commercial reasons (see **afforestation**). It is anomalous that the larch, a species common to many coniferous forests, is a deciduous conifer.

conservation   The preservation and management of the natural **environment**. In its strictest form, conservation may mean total protection of endangered species and habitats, as in nature reserves. In some cases, conservation of the man-made environment, e.g. ancient buildings, is undertaken.

continental climate   The climate at the centre of large landmasses, typified by a large annual range in temperature, with precipitation most likely in the summer.

continental drift   The theory that the Earth's continents move gradually over a layer of semi-molten rock underneath the Earth's **crust**. It is thought that the present-day pattern of continents is derived from the supercontinent **Pangaea** which existed approximately 200 million years ago. See also **Gondwanaland, Laurasia** and **plate tectonics**.

continental shelf   The seabed bordering the continents, which is covered by shallow water – usually of less than 200 metres. Along some coastlines the continental shelf is so narrow it is almost absent. Sunlight easily penetrates the seas of continental shelves, therefore plankton, plants and fish are abundant there. The British Isles have a broad continental shelf.

contour   A line drawn on a map to join all places at the same height above sea level. Contour heights are expressed in metres on British Ordnance Survey maps (e.g. the 1:50,000 series), and the interval between contours is usually 10 m (the 1:50,000 series) or 5 m (the 1:25,000 series). Contours are generally shown in brown.

contour ploughing   A method of soil **conservation** whereby ploughing is undertaken along **contours** rather than with the slope. The effect of this strategy is to reduce the rate of runoff and thus to retain **soil** that would otherwise be washed off downslope. The risk of sheet erosion and gully erosion (see **soil erosion**) is thus reduced.

contract farming   A system in which individual farmers contract with food-processing firms to produce a specific crop. The firm will often provide seed and other inputs at the start of the season, and will purchase the entire crop at harvest time. Contract farming is common in East Anglia (or a system where farmers contract with contract farmers who will farm all the land for the farmer). See also **agribusiness**.

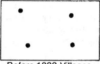

Before 1800. Villages

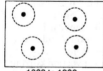

1800 to 1900.
Rapid industrialization
and urbanization

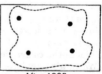

After 1900.
Towns merge into a
continuous urban
area – a conurbation

----- limit of built-up area

Conurbation The stages of growth.

conurbation   A continuous built-up urban area
formed by the merging of several formerly
separate towns or cities. Hence, for example, the
West Yorkshire conurbation includes Leeds,
Bradford, Wakefield, Huddersfield, Halifax and
many smaller towns. Several stages can be identified in the growth of conurbations.

Evidence of these processes can be observed,
for example, on the Yorkshire coalfield where
small rural **settlements** grew dramatically
during the 19th century as steam power gave
rise to the expansion of mining and the
manufacturing industry. 20th-century **urban
sprawl** has led to the merging of towns. See fig.
on previous page.

convection   The transfer of energy through a
fluid (e.g. air, water, molten material) by
movement within the fluid. An increase in the
temperature of a fluid causes an increase in its
volume and a decrease in its density. This sets up
a convection current as hotter fluid rises and
displaces the cooler, denser fluid.

Warm air circulates by convection currents in
the **atmosphere**. The same process causes
movement of liquid rock in the upper layers of
the Earth's **mantle** and is responsible for the
welling up of **magma** at a constructive plate
boundary and for the subduction of ocean crust
at a destructive plate boundary, and thus for the

movement of crustal plates (see **plate tectonics**).

cooperative A system whereby individuals pool their **resources** in order to optimize individual gains.

A farming cooperative would, for example, enable farmers to have the use of machinery which could not normally be afforded by individuals. Cooperative systems exist in many developing countries, for example there is the Ujamaa system in Tanzania.

Cooperatives may also exist in developing countries for small-scale manufacturing projects, such as the production of handicrafts for sale to tourists and for export to the developed world.

core 1. In **physical geography**, the core is the innermost zone of the Earth. It is probably solid at the centre, and composed of iron and nickel, at a very high temperature.

2. In **human geography**, a **central place** or central region, usually the centre of economic and political activity in a region or nation. John Friedman's *core / periphery model* identifies relationships between growing and stagnant regions during the process of economic development as follows:

(a) *Preindustrial phase:* independent villages

Core Independent villages.

with local **spheres of influence**.

(b) *Industrializing phase:* a core region emerges, based on a growing industrial urban centre – the rest of the nation remains an undeveloped **periphery**, supplying labour, food and other **resources** to the core.

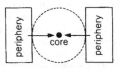

Core Industrializing phase.

(c) *Fully developed phase:* regional urban centres develop to spread the benefits of economic and social progress to the former peripheries.

This is a highly simplified version of a complex model of regional development.

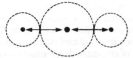

Core Fully developed phase.

corrasion   The abrasive action of an agent of **erosion** (rivers, ice, the sea) caused by its **load**, for example the pebbles and boulders carried along by a river wear away the channel bed and the river bank. Compare with **hydraulic action**.

corrie, cirque or **cwm**   A bowl-shaped hollow on a mountainside in a glaciated region; the area where a valley **glacier** originates. In glacial times the corrie contains an icefield, which in cross section appears as follows:

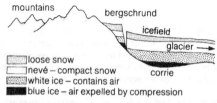

mountains   bergschrund

icefield

glacier →

corrie

☐ loose snow
☐ nevé – compact snow
☐ white ice – contains air
■ blue ice – air expelled by compression

Corrie A corrie in glacial times.

The shape of the corrie is determined by the rotational erosive force of ice as the glacier moves downslope:

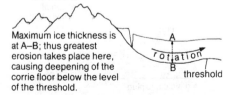

Maximum ice thickness is at A–B; thus greatest erosion takes place here, causing deepening of the corrie floor below the level of the threshold.

A
rotation
B
threshold

Corrie Erosion of a corrie.

corrosion   **Erosion** by solution action, such as the dissolving of **limestone** by running water.

counter-urbanization   The movement of population in economically developed countries from the cities to the rural areas. People move seeking a better quality of life, even though it frequently means commuting to the cities for work.

crag   A rocky outcrop on a valley side formed, for example, when a **truncated spur** exists in a glaciated valley. Crags are exposed to **weathering**, especially by **nivation**, and the resulting debris accumulates as **scree** beneath the crag.

crag and tail   A feature of lowland **glaciation**, where a resistant rock outcrop withstands **erosion** by a **glacier** and remains as a feature after the **Ice Age**. Rocks of volcanic or metamorphic

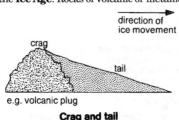

Crag and tail

origin are likely to produce such a feature. As the ice advances over the crag, material will be eroded from the face and sides and will be deposited as a mass of boulder clay and debris on the leeward side, thus producing a 'tail'. An example of a crag and tail is Castle Rock and the Royal Mile, Edinburgh.

crater lake A lake occupying a **caldera** of a dormant or extinct **volcano**.

crevasse A crack or fissure in a **glacier** resulting from the stressing and fracturing of ice at a change in **gradient** or valley shape. Crevasses

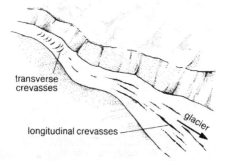

transverse crevasses

longitudinal crevasses

glacier

Crevasse Location and direction of crevasses in a glacier.

range from small surface cracks to major fractures many metres in depth, and may occur at any angle although two main orientations are recognized: transverse and longitudinal. Transverse crevasses occur where there is a change of gradient in the valley floor; longitudinal where the valley becomes wider and the ice mass stretches to occupy the broader space.

cross section A drawing of a vertical section of a line of ground, deduced from a map. It depicts the **topography** of a system of **contours**. See diagram below.

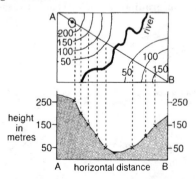

Cross section Map and corresponding cross section.

crust The outermost layer of the Earth, representing only 0.1% of the Earth's total volume. It comprises continental crust and oceanic crust, which differ from each other in age as well as in physical and chemical characteristics. The most ancient parts of the continental crust are 4000 million years old, whereas the oldest oceanic crust has an age of less than 200 million years. The oceanic crust is much thinner than the continental crust (5-10 km compared with 25-90 km). The crust, together with the uppermost layer of the **mantle**, is also known as the *lithosphere*. See also **plate tectonics**.

culvert An artificial drainage channel for transporting water quickly from place to place. Culverting schemes are important in areas prone to flooding. Concrete-lined culverts, with less frictional resistance than a river channel, allow water to flow more freely, thus lessening the risk of flooding during periods of high rainfall.

cumecs *abbreviation for* 'cubic metres per second'. See **discharge**.

cumulonimbus A heavy dark **cloud** of great vertical height. It is the typical thunderstorm cloud, producing heavy showers of rain, snow or hail. Such clouds form where intense solar radiation causes vigorous convection.

cumulus A large **cloud** (smaller than a **cumulonimbus**) with a 'cauliflower' head and almost horizontal base. It is indicative of fair or, at worst, showery **weather** in generally sunny conditions.

cut-off See **oxbow lake**.

cyclone See **hurricane**.

dairying A **pastoral farming** system in which dairy cows produce milk that is used by itself or used to produce dairy products such as cheese, butter, cream and yoghurt.

death rate The number of deaths per 1000 people in a population per year.

deciduous woodland Trees which are generally of broad-leaved rather than **coniferous** habit, and which shed their leaves during the cold season. The larch is a deciduous conifer and is thus the exception to the **evergreen** norm in such trees

deflation The removal of loose sand by wind **erosion** in desert regions. It often exposes a bare rock surface beneath.

deforestation The practice of clearing trees.

Much deforestation is a result of development pressures, e.g. trees are cut down to provide land for agriculture and industry. Deforestation in some **Third World** countries had led to severe **soil erosion**, a consequence of which has been **desertification** and eventually famine. It is thought that many of the famines in African countries during recent years have been partly caused by deforestation.

The felling of huge areas of **tropical rainforest** is causing a change in the balance of oxygen to carbon dioxide in the **atmosphere** (see **global warming**).

delta A fan-shaped mass consisting of the deposited **load** of a river where it enters the sea. A delta only forms where the river deposits material at a faster rate than can be removed by coastal currents. While deltas may take almost any shape and size, three types are generally recognized, as shown in figure overleaf.

demographic transition theory A model of **population change** which suggests the following pattern of changes over time:
1) *Preindustrial societies:* a high **death rate** due to factors such as low levels of technology and vulnerability to natural disaster; a high **birth rate** to ensure survival of the population.
2) *Industializing societies:* improvements in e.g.

Arcuate delta
e.g. Nile
Note bifurcation of
river into
distributaries in delta.

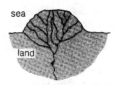

Bird's foot delta
e.g. Mississippi

Estuarine delta
e.g. Amazon

Delta

nutrition, **communications** and medical
facilities, have a quick effect on the death rate
which falls rapidly; the birth rate remains high.
3) *Urban/industrial societies:* in highly

developed economies the majority of the population is urban and in paid employment; as a result birth rate falls. Small families are likely partly as a result of a much more mobile population and partly because large families are no longer needed to provide the labour required in pre-industrial, agricultural societies.

The theory can be seen as the history of one nation's population evolution, or a classification applicable to a number of different nations at various stages in the development process. Although the model poses a very generalized and simplified form of reality, it does emphasize the essential link between **population change**, economic development and **urbanization**.

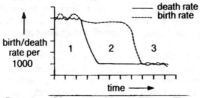

Demographic transition theory The demographic transition model.

denudation The wearing away of the Earth's surface by the processes of **weathering** and **erosion**.

depopulation An actual decrease in the population of any given area, frequently caused by economic migration to other areas.

deposition The laying down of **sediments** resulting from **denudation**.

depressed region An area of economic stagnation or decline, characterized by high unemployment, out-migration, and declining private and public investment. Depressed regions are often sites of traditional **heavy industry** such as steel-making and engineering; as the **resource** base for such activities contracts, and as demand for their products falls, so the regional economy decays. Parts of northeast England, Merseyside and Clydeside are depressed regions in Britain. Central government attempts to revitalize depressed areas through **development area** policy.

depression An area of low atmospheric pressure occurring where warm and cold air masses come into contact, for example, in the case of the northern hemisphere, along the north polar front (30°–60°N) where prevailing southwesterly winds (bringing moist, warm tropical air northwards) meet prevailing northeasterly winds (bringing cold polar air southwards). The passage of a depression is marked by thickening cloud, rain, a period of dull and drizzly weather and then clearing skies with showers. A depression develops as follows:

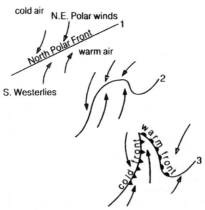

Depression The development of a depression.

desert An area where all forms of **precipitation** are so low that very little, if anything, can grow. On average a desert gets less than 250 mm of rainfall a year, though some deserts can receive more than 500 mm a year. However, due to rapid run-off and high rates of **evaporation**, this rainwater is unable to support a great deal of life.

Deserts can be broadly divided into three types, depending upon average temperatures:

(a) *hot deserts:* occur in tropical latitudes in regions of high pressure where air is sinking and therefore making rainfall unlikely. See **cloud**. The predominant westerly winds blow across cold currents before reaching land, and therefore most rainfall is deposited offshore. The daytime temperatures in hot deserts can reach 50°C dropping to −10°C at night, i.e. the diurnal range of temperature is very high.

(b) *temperate deserts:* occur in mid-latitudes in areas of high pressure. They are far inland, so moisture-bearing winds rarely deposit rainfall in these areas. The summer temperatures in these deserts may reach 20°C and will drop to below −20°C during the winter.

(c) *cold deserts:* occur in the northern latitudes, again in areas of high pressure. Very low temperatures throughout the year mean the air

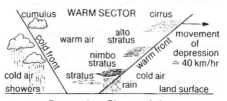

Depression Characteristics.

is unable to hold much moisture. Summer temperatures may reach 15°C whilst winter temperatures plunge to below –60°C. See also **desertification**.

desertification The encroachment of **desert** conditions into areas which were once productive. Desertification can be due partly to climatic change, i.e. a move towards a drier climate in some parts of the world (possibly due to **global warming**), though human activity has also played a part through bad farming practices. The problem is particularly acute along the southern margins of the Sahara desert in the *Sahel region* between Mali and Mauritania in the west, and Ethiopia and Somalia in the east. Low rainfall for several years plus poor management of the land have caused the edge of the Sahara desert to extend southwards.

Although desertification is particularly serious in the Sahel, there are several other parts of the world where it is also a problem.

Improved farming practices – irrigation schemes, controlled grazing and **afforestation** – can slow down the process of desertification.

developing countries A collective term for those nations in Africa, Asia and Latin America which are undergoing the complex processes of modernization, **industrialization** and **urbanization**. There is great political and social variation between countries loosely grouped as 'developing', and such terms should be used with caution. See also **Third World**.

development area   In Britain, a region designated by government as in need of special assistance for economic reconstruction. See **depressed region**.

development process   The sequence of events by which a nation moves from a predominantly subsistence agricultural economy to one based on **commercial agriculture**, industry and a highly urbanized society.

dew point   The temperature at which the **atmosphere**, being cooled, becomes saturated with water vapour. This vapour is then deposited as drops of dew.

differential erosion   The unequal **erosion** of interbedded hard and soft rocks, the softer **strata** being worn away more quickly. See, for example, **bay**.

dip   The angle of inclination of **strata** from the horizontal.

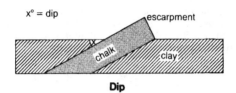

Dip

dip slope The gentler of the two slopes on either side of an escarpment crest; the dip slope inclines in the direction of the dipping **strata**; the steep slope in front of the crest is the **scarp slope**.

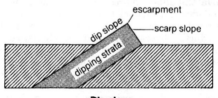

Dip slope

discharge The volume of run-off in the channels of a river **basin**.

discordant coastline A coastline that is at right angles to the mountains and valleys immediately inland. A rise in sea level or a sinking of the land will cause the valleys to be flooded. A flooded river valley is known as a **ria**, whilst a flooded glaciated valley is known as a **fiord**. Compare **concordant coastline**.

sea

land

Discordant coastline

diversification   A broadening of an agricultural or industrial product range, in order to reduce dependence on a single, perhaps vulnerable, product. Diversification in **agriculture** is ecologically sound since a more varied **ecosystem** will have a healthier pest-predator complex – uniform wheatfields, for example, are susceptible to explosions of plant-specific pests which require expensive (and perhaps pollutive) artificial control. Such **monoculture** also progressively drains the **soil** of specific nutrients. Diversification is economically sound as an insurance against falling markets for a single product.

doldrums   An equatorial belt of low atmospheric

pressure where the **trade winds** converge. Winds are light and variable but the strong upward movement of air caused by this convergence produces frequent thunderstorms and heavy rains.

dormitory settlement   A village located beyond the edge of a city but inhabited by residents who work in that city (see **commuter zone**). The populations of dormitory **settlements** are disproportionately large for the number of goods and **services** available within them (see **central place theory**).

drainage   The removal of water from the land surface by processes such as streamflow and infiltration. Drainage can be hastened artificially by the laying of pipes and culverts.

drainage basin   See **river basin**.

drift   Material transported and deposited by glacial action on the Earth's surface. See also **boulder clay**.

drift mine   A system of mining in which an inclined plane gives access to the ore. In areas of the Yorkshire coalfield inclined roadways lead to the coalface, allowing free access for plant and machinery. Compare **shaft mine**.

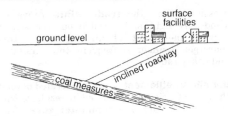

Drift mine Access to coal seam.

dry valley or **coombe**   A feature of **limestone** and **chalk** country, where valleys have been eroded in what are today dry landscapes. Such valleys may date from a period of more moist **climate**, or from the end of the last **glaciation** when *periglacial* conditions, specifically a frozen subsoil and bedrock, sealed the otherwise **permeable** limestone and thus surface streams existed.

dyke   1. An artificial **drainage** channel.

   2. An artificial bank built to protect low-lying land from flooding.

   3. A vertical or semi-vertical igneous intrusion occurring where a stream of **magma** (e.g. from a **batholith**) moves through a line of weakness in the surrounding **rock**.

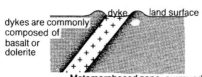

dykes are commonly composed of basalt or dolerite

Metamorphosed zone: surrounding rocks close to intrusion are 'baked'.

Dyke Cross section of eroded dyke, showing how metamorphic margins, harder than dyke or surrounding rocks, resist erosion.

earthquake A movement or tremor of the Earth's crust. Earthquakes are associated with plate boundaries (see **plate tectonics**) and

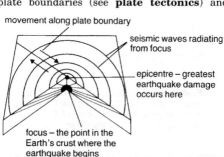

movement along plate boundary

seismic waves radiating from focus

epicentre – greatest earthquake damage occurs here

focus – the point in the Earth's crust where the earthquake begins

Earthquake

especially with subduction zones, where one plate plunges beneath another. Here the crust is subjected to tremendous stress. The rocks are forced to bend, and eventually the stress is so great that the rocks 'snap' along a **fault** line. This energy is released as *seismic waves*, originating from the *focus*, i.e. where the earthquake begins.

eastings   The first element of a **grid reference**, as in the following example:

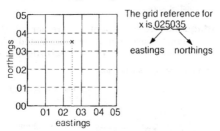

The grid reference for x is 025035

eastings   northings

Eastings The bottom left corner of the map is taken as the origin; eastings are read towards the right edge of the map, northings towards the top edge.

ecology   The study of living things, their inter-relationships and their relationships with the

environment.

economies of scale The savings made in industry through mass production, automation and integrated processes. The unit cost of products falls as the quantity manufactured increases; investment in sophisticated manufacturing processes can further reduce unit costs.

ecosystem A natural system comprising living organisms and their **environment**. The concept can be applied at the global scale or in the context of a smaller defined environment, e.g. a woodland, a pond or a marsh. Whatever the scale, the principle of the ecosystem is constant: all elements are intricately linked by flows of energy and nutrients, and a change in one element will have effects on the rest of the system.

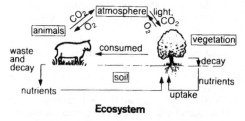

Ecosystem

emigration The movement of population out of

a given area or country.

employment structure   The distribution of the workforce between the **primary, secondary, tertiary** and **quaternary** sectors of the economy. Primary employment is in **agriculture**, mining, forestry and fishing; secondary in manufacturing; tertiary in the retail, service and administration category; quaternary in information and expertise. One way of looking at the level of development of a nation is to examine its employment structure. Most employment in the **Third World** is primary, while that in the more developed nations is tertiary, with an increasing number of people being employed in the quaternary sector.

enterprise zone   An area, usually in the inner city, where special grants and facilities are made available for reconstruction and development.

environment   Physical   surroundings: **soil**, vegetation, wildlife and the **atmosphere**. Human impact on the environment is a major concern of geographers, especially as human interference can often create problems – for example in causing **pollution**, **soil erosion**, the extinction of species and spread of urban areas. In a broader sense, the term environment is also used to describe the social as well as the physical surroundings of people, such as culture, language, traditions and political systems.

erosion The wearing away of the Earth's surface by running water (rivers and streams), moving ice (**glaciers**), the sea and the wind. These are called the *agents* of erosion.

erratic A boulder of a certain rock type resting on a surface of different geology. For example, at Norber Blocks in Yorkshire blocks of **granite** rest on a surface of carboniferous **limestone**.

The explanation is glacial: the erratics were picked up by **glaciers** moving southwards over Britain and deposited at the point where the ice melted. Erratic blocks of this kind may be found at considerable distances from their origin.

escarpment A ridge of high ground as, for example, the **chalk** escarpments of southern England (the Downs and the Chilterns).

esker A low, winding ridge of pebbles and finer **sediment** on a glaciated lowland. An esker marks the course of a subglacial meltwater stream; as the stream flows it deposits sediment from the glacier along its course, and this is left behind as a landscape feature after the end of the glacial period.

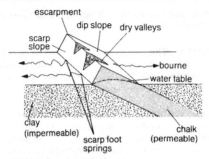

Escarpment A simplified block diagram of a chalk escarpment.

estuary The broad mouth of a river where it enters the sea. An estuary forms where opposite conditions to those favourable for **delta** formation exist: deep water offshore, strong marine currents and a smaller **sediment** load (perhaps due to **deposition** upstream, e.g. in a lake).

ethnic group A group of people with a common identity such as culture, religion or skin colour.

EU (European Union) Formerly known as the European Community (EC) and the European Economic Community (EEC), this is a group of fifteen European countries, (Austria, Belgium, Denmark, Finland, France, Germany, Greece, Republic of Ireland, Italy, Luxembourg, Netherlands, Portugal, Spain, Sweden and the United Kingdom) which have agreed to have close political and economic links with each other.

eutrophication   The process whereby an area of water becomes rich in nutrients, causing excessive growth of plants (usually algae) in the springtime. When they decompose, the algae use up the dissolved oxygen in the water, so that wildlife in the pond or lake dies from lack of oxygen.

Eutrophication can be a problem in agricultural areas where fertilizers have leached into the waterways.

evaporation   The process whereby a substance changes from a liquid to a vapour. Heat from the sun evaporates water from seas, lakes, rivers, etc., and this process produces water vapour in the **atmosphere**.

evaporite   A type of **sedimentary** rock formed where salts are precipitated by evaporation in hot climates. Typical evaporites are gypsum and halite (rock salt), which are formed around the margins of lakes in **basins of internal drainage**, e.g. the Dead Sea and the Sea of Galilee.

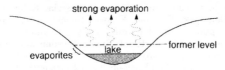

Evaporites The position of evaporites on a lake margin.

evergreen A vegetation type in which leaves are continuously present. Compare **deciduous woodland**.

evapotranspiration The return of water vapour to the **atmosphere** by evaporation from land and water surfaces and the **transpiration** of vegetation.

exfoliation A form of **weathering** whereby the outer layers of a **rock** or boulder shear off due to the alternate expansion and contraction produced by diurnal heating and cooling. Such a process is

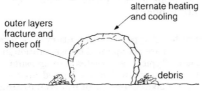

Exfoliation

especially active in **desert** regions where, due to clear skies, night temperatures drop considerably below daytime peaks. Isolated boulders may be surrounded by exfoliation debris.

expanded town An existing town where growth is planned in order to accommodate newcomers. Funding comes from local and national government. Expanded towns are favoured as being less costly than **new towns**. Examples in England include Swindon and Basingstoke.

exports Goods and services sold to a foreign country (compare **imports**).

exposed coalfield See **concealed coalfield**.

exponential growth A rapid form of growth,

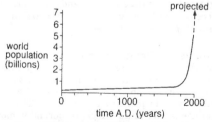

Exponential growth A line graph to show world population growth.

for example where a population doubles in every unit of time: 2, 4, 8, 16, 32. This can be contrasted with *arithmetic growth*, for example 2, 4, 6, 8, 10 where the growth is the same in every unit of time.

extensive farming   A system of **agriculture** in which relatively small amounts of capital or labour investment are applied to relatively large areas of land. For example, sheep ranching is an extensive form of farming, and yields per unit area are low. Extensive farming usually occurs at the *margin* of the agricultural system – at a great distance from the market, or on poor land of limited potential. See also **Von Thünen theory**.

external   processes   Landscape-forming processes such as **weather** and **erosion**, in contrast to internal processes.

extreme   climate   A climate that is characterized by large ranges of temperature and sometimes of rainfall. In central Asia for example, hot summers (average 21°C) alternate with very cold winters (average −45°C). Thus the average temperature range is normally more than 60°C. Most rainfall occurs in summer. Compare **temperate climate, maritime climate**.

false colour image   See **satellite image**.

fault   A fracture in the Earth's crust either side of which the **rocks** have been relatively displaced. The scale of faulting can vary considerably, from a small surface crack to a fracture of regional extent. Faulting occurs in response to stress in the Earth's crust; the release of this stress in fault movement is experienced as an **earthquake**. See also **rift valley**.

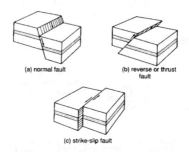

(a) normal fault

(b) reverse or thrust fault

(c) strike-slip fault

Fault The main types. Arrows show the direction of movement.

fell   Upland rough grazing in a **hill farming** system, for example in the English Lake District.

Fell land is sometimes **common land**, i.e. not in the ownership of a single individual or institution. Sheep are grazed on the fells in the summer months and brought down to lower pasture during the winter.

field sketch   A sketch of a landform or landscape made in the field. The value of a field sketch is that it forces the observer to examine carefully the features which are in view. Field sketches are usually simple, yet pick out key geographical features. Careful labelling is an integral part of any field sketch.

fjord   A deep, generally straight, inlet of the sea

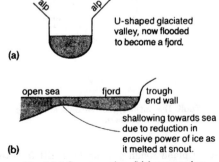

(a) U-shaped glaciated valley, now flooded to become a fjord.

(b) open sea — fjord — trough end wall — shallowing towards sea due to reduction in erosive power of ice as it melted at snout.

Fjord (a) Cross section. (b) Long section.

along a glaciated coast. A fjord is a glaciated valley which has been submerged either by a post-glacial rise in sea level or a subsidence of the land.

flash flood A sudden increase in river **discharge** and overland flow due to a violent rainstorm in the upper river **basin**.

flood plain The broad, flat valley floor of the lower course of a river, levelled by annual flooding and by the lateral and downstream movement of **meanders**.

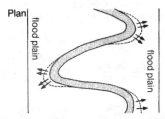

Flood plain Due to concentration of erosion on the outer banks of meanders, these 'migrate' across and along the flood plain.

flow line A diagram showing volumes of movement, e.g. of people, goods or information between places. The width of the flow line is proportional to the amount of movement, for example in portraying commuter flows into an urban centre from surrounding towns and villages:

Flow line Commuter flows into a city.

fodder crop A crop grown for animal feed, either for direct feeding, e.g. turnips, or for making **silage**, as with grass grown for hay.

fold A bending or buckling of once horizontal

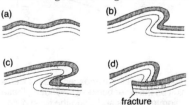

Fold (a) Syncline and anticline. (b) Overfold. (c) Recumbent fold. (d) Overthrust fold or nappe; here the rock has been fractured.

rock **strata**. Many folds are the result of rocks being crumpled at plate boundaries (see **plate tectonics**), though **earthquakes** can also cause rocks to fold, as can **igneous intrusions**. The flat plain of southern England has been folded into a series of anticlines (arch-shaped) and synclines (trough-shaped).

fold mountains Mountains that have been formed by large-scale and complex folding. Studies of typical fold mountains (the Himalayas, Andes, Alps and Rockies) indicate that folding has taken place deep inside the Earth's **crust** and upper **mantle** as well as in the upper layers of the crust.

Such folding is the result of compression of the continental crust when plates converge (see **plate tectonics**). When continental crust and oceanic crust converge, the oceanic crust is carried downwards beneath the continental plate (*subduction*), and the leading edge of the continental crust is compressed and folded upwards to form a mountain range parallel with the coast, such as the Andes. When two continental plates collide, folding and uplifting occurs at both leading edges as the plates fuse together. The Alps and the Himalayas were formed in this way.

food chain The feeding succession, starting

with *producers*, i.e. green plants, and leading through to various *consumers*, i.e. animals.

In practice single food chains do not exist. Several chains interwoven form a food web. See **Ecosystem**.

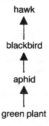

Food chain

footloose industry Any manufacturing industry with no specific requirements or preconditions for its location. Such industries are neither resource- or market-orientated and as such can enjoy a freedom to locate in a wide variety of areas. Footloose industries tend to be found in modern consumer societies in which motorways provide good transport mobility and where there is an industrial trend towards assembling products from components shipped in from a wide variety of sources. Most **light**

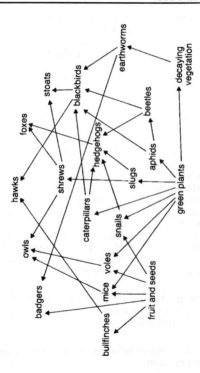

Food chain A food web.

industries are footloose as they do not have to be located close to **raw materials** or power supplies. The location of footloose industries contrasts with that of traditional **heavy industries** which were restricted to specific areas. Iron and steel plants, for example, were located close to supplies of coal and iron ore.

fossil fuel Any naturally occurring carbon or hydrocarbon fuel, notably coal, oil, peat and natural gas. These fuels have been formed by decomposed prehistoric organisms.

The burning of fossil fuels, particularly since the **industrial revolution**, has led to problems of **acid rain** in many countries of the world and to increase in levels of carbon dioxide (CO_2) in the atmosphere, a factor in the **greenhouse effect**. Most governments are committed to stabilizing or reducing CO_2 emission from burning fossil fuels. Fossil fuels are nonrenewable energy resources (see **renewable resources, nonrenewable resources**).

free trade The movement of goods and services between countries without any restrictions (such as quotas, tariffs or taxation) being imposed.

front A boundary between two air masses. See also **depression**.

functions The term used for goods and **services** available in a **central place**; in general the number and variety of functions offered increase with the size of the **settlement**. Everyday or convenience (low order) functions are available in small settlements, while these plus specialized or durable (high order) functions are available in large settlements. Small settlements may offer only two or three functions; large settlements many hundreds. See also **central place theory**.

GDP (Gross Domestic Product) The total value of the goods and services produced annually by a nation.

gentrification The process whereby old rundown houses, often in **inner city** areas, are improved and modernized by outsiders who have higher incomes than the local population. Gentrification can change the social structure of a community; as house prices rise, so the local population are no longer able to afford to rent or purchase the modernized properties.

geosyncline A basin (a large **syncline**) in which thick marine sediments have accumulated.

geothermal energy A method of producing power from heat contained in the lower layers of the Earth's **crust**. New Zealand and Iceland both use superheated water or steam from geysers and volcanic **springs** to heat buildings and for hothouse cultivation and also to drive steam turbines to generate electricity. In Britain, experiments in Southampton have successfully used heated water from rocks below the city to heat shops and offices in the city centre. In the long term, scientists are hoping to be able to tap heat from the granite rocks in southwest England. Geothermal energy is an example of a renewable resource of energy (see **renewable resources, nonrenewable resources**). It is relatively pollution free but hot mineral waters not used have to be disposed of carefully to avoid polluting surface drainage systems.

glacial advance   The extension of **ice sheets** and **glaciers** to lower altitudes to cover large areas. This is caused by a cooling of the **climate**.

glacial retreat   A reduction in the area covered by **ice sheets** and **glaciers**, caused by a warming of the **climate**. Compare **glacial advance**.

glaciation   A period of cold **climate** during which time **ice sheets** and **glaciers** are the dominant forces of **denudation**.

The last glaciation ended about 10 000 years ago, and much of Britain's landscape shows evidence of the effects of ice (north of a line drawn approximately between the Thames and the Severn).

glacier   A body of ice occupying a valley and originating in a **corrie** or icefield. A glacier moves at a rate of several metres per day, the precise speed depending upon climatic and **topographic** conditions in the area in question. The Mer de Glace and Aletsch glaciers in the Alps are present-day examples.

globalization   The process that enables financial markets and companies to operate internationally (as a result of deregulation and improved communications). **Transnational corporations** now locate their manufacturing in places that best serve their global market at the lowest cost.

global warming or **greenhouse effect**   The warming of the Earth's atmosphere caused by an excess of carbon dioxide, which acts like a blanket, preventing the natural escape of heat. This situation has been developing over the last 150 years because of (a) the burning of **fossil fuels**, which releases vast amounts of carbon dioxide into the **atmosphere**, and (b) **deforestation**, which results in fewer trees being available to take up carbon dioxide (see **photosynthesis**).
A global rise in temperature is likely to have

many consequences. For example, parts of the polar ice sheet would melt, raising sea level, which would result in the flooding of low-lying coastal areas throughout the world. This would cause loss of life, and in addition, the loss of farmland and food supplies, housing and industry. It would damage transport networks and power supplies, and would pollute inland water supplies with salt water. In other parts of the world, rising temperatures would cause the development of new areas of desert, making food production difficult in formerly fertile areas.

The process of global warming could be slowed down by reduction of the use of fossil fuels, replanting of trees and stricter controls on the destruction of the world's forests.

GNP (Gross National Product)   The total value of the goods and services produced annually by a nation, plus net property income from abroad. GNP per capita is the GNP divided by the total population of the country. It is a crude measure as it does not take account of the purchasing power of money, the income distribution in the country, or the real income in developing countries where many farmers grow food for subsistence rather than cash.

Gondwanaland   The   southern   hemisphere

supercontinent, consisting of the present South America, Africa, India, Australasia and Antarctica, that split from **Pangaea** *c* 200 million years ago. Gondwanaland is part of the theory of **continental drift**. See also **plate tectonics**.

graded profile The long profile of a river's course that would exist after all irregularities have been removed by **erosion**.

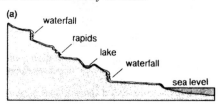

(a)
waterfall
rapids
lake
waterfall
sea level

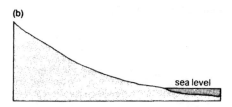

(b)
sea level

Graded profile (a) Ungraded profile. (b) Graded profile.

gradient   1. The measure of steepness of a line or slope. In mapwork the average gradient between two points can be calculated as:

$$\frac{\text{difference in altitude}}{\text{distance apart}}$$

Gradient

Gradient between A & B:

$$\frac{300 \text{ m} - 50 \text{ m}}{800 \text{ m}}$$

$$= \frac{250 \text{ m}}{800 \text{ m}} = \frac{1}{3 \cdot 2} \qquad \begin{array}{l}\textbf{Expressed} \\ \textbf{as } 1 : 3 \cdot 2\end{array}$$

Interpretation: 1 m of ascent for every 3.2 m of horizontal equivalent.

2. The measure of change in a property such as density. In **human geography** gradients are found in, for example, **population density**, land values and **settlement** ranking.

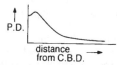

Gradient A typical graph showing the urban population density gradient. It indicates a general decline in population density with increasing distance from city centres.

granite An **igneous rock** having large crystals due to slow cooling at depth in the Earth's **crust**.

Granite is a common constituent of **batholiths**, and is mainly composed of **quartz**, mica and feldspar minerals.

green belt An area of land, usually around the outskirts of a town or city on which building and other developments are restricted by legislation. The purpose of such planning law is to attempt to preserve open space and relatively rural **environments** which would otherwise be lost with the advance of **urban sprawl**.

greenfield site   A development site for industry, retailing or housing that has previously been used only for agriculture or recreation. Such sites are frequently in the **green belt**.

greenhouse effect   See **global warming**.

green revolution   The introduction of high yielding crops of rice and wheat into developing countries. These crops require fertilizers and can only be used by farmers who have appropriate funds. They have helped to increase food production.

grid reference   A method for specifying position on a map. See **eastings**.

groundwater   Water held in the bedrock of a region, having percolated through the **soil** from the surface. Such water is an important **resource** in areas where **surface runoff** is limited or absent.

gryke   An enlarged joint between blocks of **limestone** (**clints**), especially in a **limestone pavement**. Grykes are progressively enlarged by the solution effect of rainwater.

habitat   A preferred location for particular species of plants and animals to live and reproduce. A habitat may be small, such as a rock-cranny, or large, such as a **tropical rain forest**.

hanging valley   A tributary valley entering a main valley at a much higher level because of deepening of the main valley, especially by glacial erosion. During **glaciation**, vertical erosion is much greater in the main valley than in smaller tributary valleys. Small tributary valleys will only have small glaciers in them or no glaciers at all. Once the ice has retreated the floor of the main valley lies far below the tributary valleys which are therefore called 'hanging valleys'.

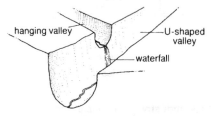

Hanging valley

Harris and Ullman model   See **multiple nuclei model**.

HDI (human development index) A measure created by the United Nations to assess the economic and social development of a country. It combines information on life expectancy, education, and the purchasing power of incomes. The index goes from 0 to 1, the highest score representing the most developed.

headland A promontory of resistant **rock** along the coastline. See **bay**.

heave A horizontal displacement of **strata** at a fault.

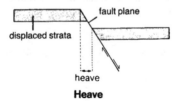

Heave

heavy industry Traditional industries using bulky **resources**, e.g. coal mining, iron and steel-making, chemicals and engineering.

high-technology approach An approach to the **development process** which stresses the role of capital and sophisticated technology. It is argued that investment in large-scale **resource** management schemes (e.g. dams for **hydroelectric power**), and in the industrial sector in general, is the surest way to hasten national development. In Brazil, for example, development priorities have been identified in this way.

Very often the capital for high-technology schemes in developing countries is provided by developed countries. Contrast this with the **intermediate technology** (and **appropriate technology**) approach.

hill farming A system of **agriculture** where sheep are grazed (and to a lesser extent cattle) on upland rough pasture.

In Britain hill farming occurs on **marginal land** in upland areas such as the Lake District, Snowdonia and the Scottish Highlands. The typical hill farm comprises three zones: the *inbye, intake* and **fell**. The inbye is valley-bottom land, immediately surrounding the farm buildings; it is walled or fenced and may be cultivated for **fodder crops** and sown pasture. The intake extends up the lower slopes of the surrounding **fells** and is an area of sheltered pasture for winter grazing and for lambing. The fell is an extensive area of upland rough grazing to which several farmers may have right of access.

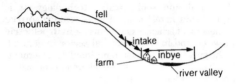

Hill farming A typical hill farm.

hinterland An area inland from a port, and defined by the limit of the port's **sphere of influence**, for example:

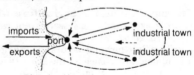

- - - - hinterland boundary
- - - - flows of raw materials – food, ores etc.
——► flows of finished products
- - - ► flows of agricultural produce

Hinterland

histograph A graph for showing values of classed data as the areas of bars. See diagram opposite. Compare **bar graph**.

home region The area around a person's home. This might be on a small scale, e.g. the area

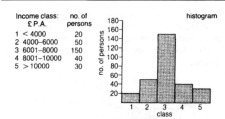

Income class: £ P.A.	no. of persons
1 < 4000	20
2 4000–6000	50
3 6001–8000	150
4 8001–10000	40
5 > 10000	30

Histogram

within a short travelling distance, or on a broader scale, several counties. For example, the home region for a person living in London could be regarded as southeast England.

honeypot A key site in a tourism area that particulary attracts visitors so pulling visitors into an area they might otherwise not visit.

horizon The distinct layers found in the **soil profile**; usually three horizons are identified – A, B, C as follows:

The A horizon or **topsoil** contains humus and other vegetable debris. The B horizon or subsoil contains a larger proportion of inorganic material, and receives minerals washed down from the topsoil by the process of **leaching**. The division between the B and C horizons is marked by a zone of decaying bedrock. In reality there will rarely be sharp divisions between zones; the A and B horizons, for example, may be mixed by the activity of worms, burrowing animals or root growth. See also **inorganic fraction, organic fraction**.

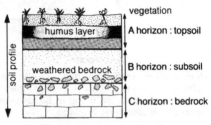

Horizon A typical soil profile.

horticulture   The growing of plants and flowers for commercial sale. It is now an international trade, for example, with orchids being grown in Southeast Asia for sale in Europe.

household   The number of people living in a single dwelling. This could be one person in a bedsit or a family of six in a house. A block of flats would be made up of several households. Information about households is required by the 10-year national **census**.

household amenities Utilities in a dwelling such as gas, electricity and running hot and cold water, which are important for everyday life. Older dwellings may have poorer amenities, such as a shared bathroom and an outside toilet, compared with many modern dwellings which have very good amenities, e.g. central heating. See also **neighbourhood amenities**.

human geography The study of people and their activities in terms of patterns and processes of population, **settlement**, economic activity and **communications**. There is no precise definition of such a broad subject, but the basic task of the human geographer is to try to explain distributions of people and their activities. Compare **physical geography**.

hunter/gatherer economy A pre-agricultural phase of development in which people survive by hunting and gathering the animal and plant **resources** of the natural **environment**. No cultivation or herding is involved. Very few hunter/ gatherer societies survive today, but the few examples we do have include the Wiama Indians and other tribes living in the Amazon rainforest.

hurricane, cyclone or **typhoon** A wind of force 12 on the **Beaufort wind scale**, i.e. one having a velocity of more than 118 km per hour.

Hurricanes can cause great damage by wind as well as from the storm waves and floods that they help to produce.

Hurricanes occur in the belt of the **trade winds** where these winds begin to diminish towards the **doldrums**. Such tropical storms usually arise at the end of summer when the ocean is warmest. This means that the warm, moist air above the ocean is likely to rise, and **condensation** will occur. This releases latent energy (i.e. the heat originally used to **evaporate** the water) as the air spirals upwards and warms the centre, or *eye*, of the storm. This warm, dry air descends, expands and sucks up more moisture, which again is carried upwards. Thus, the hurricane is 'fed' by the warm ocean, and wind speeds increase as the winds circle the eye of the storm. The hurricane will continue as long as it is fed with moisture from the sea and with heat from condensation. This explains why most hurricanes lose force and 'die' when they reach land, though severe damage can occur in coastal areas.

hydraulic action The erosive force of water alone, as distinct from **corrasion**. A river or the sea will erode partially by the sheer force of moving water and this is termed 'hydraulic action'. For example, a swift river current will undercut the outer bank of a **meander**. The force of waves breaking against a **headland** will compress the air in rock crevices and thereby cause the surrounding **rock** to shatter.

hydroelectric power The generation of electricity by turbines driven by flowing water. Hydroelectricity is most efficiently generated in rugged **topography** where a head of water can most easily be created, or on a large river where a dam can create similar conditions. Whatever the location, the principle remains the same – that water descending via conduits from an upper storage area passes through turbines and thus creates electricity.

hydrological cycle The cycling of water through sea, land and **atmosphere**.

The amount of fresh water available in the system is small: 97% of all the water on the Earth's surface is salt, and of the remaining 3% which is fresh, most is locked in **ice sheets**. See figure on next page.

hydrograph See **storm hydrograph**.

hydrosphere All the water on Earth, including that present in the **atmosphere** as well as in oceans, seas, **ice sheets**, etc.

hygrometer An instrument for measuring

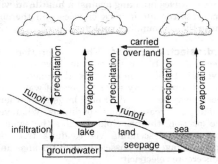

Hydrological cycle

the relative humidity of the **atmosphere**. It comprises two thermometers, one of which is kept moist by a wick inserted in a water reservoir. Evaporation from the wick reduces the temperature of the 'wet bulb' thermometer, and the difference between the dry and the wet bulb temperatures is used to calculate relative humidity from standard tables.

Ice Age A period of **glaciation** in which a cooling of **climate** leads to the development of **ice sheets, ice caps** and valley **glaciers**. The most recent Ice Age was the Quaternary glaciation, ending about 10 000 years ago.

Compare **Little Ice Age**.

ice cap A covering of permanent ice over a relatively small land mass, e.g. Iceland.

ice fall An area of fractured ice in a **glacier** where a change of **gradient** occurs.

ice sheet A covering of permanent ice over a substantial continental area such as Antarctica.

igneous rock A **rock** which originated as **magma** (molten rock) at depth in or below the Earth's **crust**. Igneous rocks are generally classified according to crystal size, colour and mineral composition; intrusive and extrusive types are also recognized.
1) **batholith:** a large body of magma intruded into the Earth's crust; this cools slowly at depth to form igneous rocks with large crystals such as **granite**.
2) **dyke:** vertical or semi-vertical sheet of igneous rock; a minor intrusion compared with a batholith. Dolerite is a common dyke rock.
3) **sill:** horizontal or semi-horizontal minor intrusion; sills and dykes exploit lines of weakness (e.g. **joints, bedding planes, faults**) in the crustal rocks.
4) **lava flow:** extrusive igneous rocks are those which reach the Earth's surface via some form of

volcanic eruption. Such rocks have small crystals due to rapid cooling on the Earth's surface. **Basalt** is a common example.

The terms volcanic, *hypabyssal* and **plutonic** are also used to describe igneous rocks: **volcanic rocks** are those which are extruded onto the surface; hypabyssal rocks are those of intrusions such as sills and dykes; and **plutonic rocks** are those of deep intrusions such as batholiths.

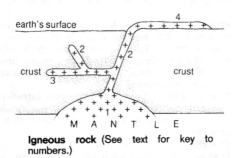

Igneous rock (See text for key to numbers.)

immigration   The movement of people into a country or region from other countries or regions.

impermeable rock   A rock that is nonporous and therefore incapable of taking in water or of allowing it to pass through between the grains. Compare **impervious**. See also **permeable rock**.

impervious rock   A nonporous rock with no cracks or fissures through which water might pass. An **impermeable** rock such as granite may be pervious due to the presence of **joints**.

imports   Goods or services bought into one country from another (compare **exports**).

inbye   See **hill farming**.

industrial estate   A purpose-built facility for the location of new industry, often situated at the edge of an urban area where land is more readily available and where there is less congestion. Motorway intersections or access points are favoured locations for industrial estates owing to the important role of road transport today. See also **business park** and **science park**.

industrial inertia   The tendency for industry to retain original locations even though such locations may no longer be optimum. Steel-making, for example, continues at Sheffield, even though the ores upon which the industry was originally based are long since worked out. The woollen **textile industry** continues to be concentrated in West Yorkshire, even though local factors such as swift streams for water power and, later, **coal** for steam power are no longer relevant. Industries retain their original locations in this way because the costs of relocating are likely to be expensive. Large amounts of capital are tied up in factory facilities; a skilled labour force may have developed in the region; economies of **agglomeration** and scale may have been established locally. For such reasons it may be cheaper to maintain the original location and to import raw materials from elsewhere. In the case of Sheffield, for example, scrap is now the major raw material for the steel industry.

industrialization The development of industry on an extensive scale, as happened during the Industrial Revolution in Western Europe. Some people identify industrialization with economic development, but for some nations other methods of development may be more appropriate. The term 'industrializing' is sometimes used to describe nations midway through the development process, and **preindustrial** and **postindustrial** nations are identified by the same measure. See also **newly industrialized countries**.

industrial location The optimum location for an industry is one where the costs of transport are minimized, with respect to both **raw materials** and finished products. The standard theory of industrial location is that of Alfred Weber's *locational triangle:* an industry may locate anywhere within the triangle, depending on the balance of transport costs relating to the three factors.

Various examples can be envisaged, e.g. an industry with bulky raw materials which are expensive to transport will locate close to factor 1.

Industrial location The locational triangle.

An industry such as printing and publishing will locate close to the market, since distribution costs are the most important consideration:

Industrial location Location to minimize distribution costs.

An industry for which all transport costs are similar will locate in the centre of the triangle. In reality of course the locational decision is considerably more complex: factors such as the availability of power and the suitability of the land will also be taken into account, as will less tangible issues such as managers' personal preferences and outlook. Government intervention, for example through **development area** policy, may also affect the locational decision. See **iron and steel industry** for an example of shifting location over time.

Industrial Revolution   The period in Britain's history from approximately 1780 to 1900, when the invention and application of industrial processes led to the establishment of the world's first urban/industrial society. The invention of the steam engine provided the key to a number of crucial industrial innovations in mining, transport and manufacture. Prior to the industrial revolution, successes within agriculture had been due to increased mechanization and efficiency, plus greater profits. This released labour to work in the new industrial cities. Throughout the 19th century large-scale **migration** to urban areas continued, and it was during this period that the major industrial concentrations were established, for example in West Yorkshire, South Lancashire, the West Midlands and Central Scotland.

infiltration The gradual movement of water into the ground. The rate at which infiltration occurs is dependent upon two main factors: whether the soil is already wet and saturated, and the porosity of the soil, i.e. whether the water can move easily between the pores of the soil. Water infiltrates quickly through a sandy soil, but slowly through a clay soil.

informal economy Employment frequently found in **Third World** cities, characterized by irregular hours and wages, limited equipment and machinery, and often operating outside the law. Examples of informal jobs would include taxi drivers who transport materials around the city for business people, shoe shiners, and traders selling goods from trays, e.g. sweets, chewing gum, fruit. In some Third World cities it is estimated that the informal sector employs between 40% and 60% of the working population.

infrastructure The basic structure of an organization or system. The infrastructure of a city includes, for example, its roads and railways, schools, factories, power and water supplies and drainage systems.

inner city The ring of buildings around the **CBD** of a town or city. These buildings are usually **terraced** houses of the early industrial period (many of which have been modernized – see **gentrification**), blocks of high and low-rise flats (to replace terraced houses which have become dilapidated), as well as offices and small **light industries**. The buildings in the inner city tend to be high density, i.e. many buildings in a relatively small area.

inorganic fraction   That proportion of the **soil** which is composed of **rock** and mineral fragments and particles deriving from the **weathering** of bedrock. Compare **organic fraction**. See also **horizon**.

intake   See **hill farming**.

integrated steelworks   An industrial site where all iron and steel manufacturing processes are carried out 'under one roof': iron manufacture, steel-making, steel rolling and processing, and alloy manufacture.

intensive farming   A system of **agriculture** where relatively large amounts of capital and/or labour are invested on relatively small areas of land. An example of intensive farming is **market gardening**, where large investment, in the form of greenhouses, fertilization, **irrigation** and heating systems, occurs on small holdings of land close to urban centres. In such a case yields per unit area are high. The small land-holdings reflect the cost of land close to market. See **Von Thünen theory** and **agribusiness**. Compare **extensive farming**.

interception   See **rainfall interception**.

interglacial   A warm period between two periods of **glaciation** and cold **climate**. The present interglacial began about 10 000 years ago.

interlocking spurs   Obstacles of hard **rock** round which a river twists and turns in a V-shaped valley. **Erosion** is pronounced on the concave banks, and this ultimately causes spurs which alternate on either side of the river and interlock.

interlocking spurs A V-shaped valley with interlocking spurs.

intermediate technology Equipment and facilities of simple, cheap, practical design directly relevant to the immediate needs of a community in a **developing country**. For example, rivers may be regulated by a series of small earthen dams to provide local flood control and water for **irrigation**. Such dams can be built with local labour and equipment, can easily be repaired and are cheap. In contrast, the **high-technology approach** of sophisticated dams for **hydroelectric power** and industrial development could involve foreign capital, equipment and expertise, and might not be appropriate for a developing country's needs. Intermediate technology, being labour intensive rather than capital intensive, does not leave the country with a big burden of debt in conditions of scarce capital. See also **appropriate technology**.

internal processes Landscape-forming processes which originate below the Earth's surface in the movements connected with **plate tectonics**: folding, faulting and **vulcanicity**.

international trade The exchange of goods and services between countries. Ideally, a country should aim to export more than it imports, thus establishing a favourable *balance of payments*. **Developing countries** tend to import more than they export, thus incurring large debts to developed countries.

intrusion A body of **igneous rock** injected into the Earth's **crust** from the **mantle** below. See **dyke, sill, batholith**.

inward investment Financial investment in a country by a **transnational corporation**. The investment may be used to develop industrial or agricultural projects in developing countries and may provide much-needed employment opportunities. The investment may also be made in developed countries, with the host government providing incentives (e.g. tax benefits or a funding contribution) to the transnational corporation. Such investments are usually made in regions requiring new economic activity. However, the outputs of such projects are usually intended for the transnational corporation's country of origin.

iron and steel industry The extraction of iron from ore and the manufacture of steel, which together are a key element in the heavy industrial structure of a nation; in Britain, the location of the iron and steel industry is an example of changing optimum location over time: the total pattern of location has three elements. Firstly, there are the coalfield locations such as Sheffield and Motherwell, dating from the early establishment of iron and steel manufacture using **coal** and blackband iron ore, found in the coal measures, as **raw materials**. Secondly, there are the orefield locations such as Scunthorpe (and formerly Corby), developed in the mid-20th century to exploit the iron ore deposits of Jurassic rocks in eastern England. By this stage the steel-making process depended less on coal, owing to the development of the electric arc furnace and, in blast furnaces, better designs which required less coal to produce iron. Thirdly, there are the most recently established coastal locations of the iron and steel industry, for example at Port Talbot in South Wales and on Teesside in northeast England. These locations reflect the dominant role today of imported ores from places such as Scandinavia and Canada. Thus the three locational elements of the overall distribution of the iron and steel industry reflect the three stages of its evolution. Steel-making in general is in decline, and several inland steelworks in the UK have closed (e.g. Consett in Co. Durham, Ravenscraig in Lanarkshire and Corby in Northamptonshire).

irrigation   A system of artificial watering of the land in order to grow crops. Irrigation is particularly important in areas of low or unreliable rainfall. Some of the oldest methods of irrigation were developed along the river Nile and consisted of simple devices such as the shadoof and the Archimedian Screw. Some of these methods are still used in the **Third World** today. The modern method of irrigation (often associated with countries of the developed world where more capital is available for investment in large-scale projects) is to build a dam across a river and create a reservoir. Water is then removed from the reservoir at times of need by a system of pipes and channels. In parts of Britain where rainfall may be unreliable, farmers use hose pipes and sprinklers, e.g. in the **arable farming** regions of East Anglia.

isobar A line joining points of equal atmospheric pressure, as on the meteorological map below.

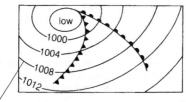

Isobar, indicating atmospheric pressure in millibars.

Isobar Section of a weather map showing isobars.

isohyet   A line on a meteorological map joining places of equal rainfall.

isotherm   A line on a meteorological map joining places of equal temperature.

joint   A vertical or semi-vertical fissure in a **sedimentary rock**, contrasted with roughly horizontal **bedding planes**. In **igneous rocks** jointing may occur as a result of contraction on cooling from the molten state. Joints should be distinguished from **faults** in that they are on a much smaller scale and there is no relative displacement of the rocks on either side of the joint. Joints, being lines of weakness, are exploited by **weathering**.

karst topography   An area of **limestone** scenery where **drainage** is predominantly subterranean. It is named after a region in Slovenia.

kettle hole   A small depression or hollow in a glacial **outwash** plain, formed when a block of ice embedded in the outwash deposits eventually melts, causing the **sediment** above to subside.

labour intensive   Denoting a system of **agriculture** or industry where labour (rather than capital) forms the major input. For example, rice cultivation in much of Southeast Asia is a labour-intensive activity with very little mechanization. As nations develop, capital tends to replace labour in certain sectors of the economy and labour intensity decreases.

laccolith An igneous **intrusion**, domed and often of considerable dimensions, caused where a body of viscous **magma** has been intruded into the **strata** of the Earth's **crust**. These strata are buckled upwards over the laccolith.

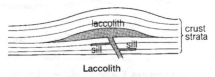

Laccolith

lagoon 1. An area of sheltered coastal water behind a **bay bar** or **tombolo**.
   2. The calm water behind a coral reef.

Lagoon A lagoon behind a bay bar.

lahar   A landslide of volcanic debris mixed with water down the sides of a volcano, caused either by heavy rain or the heat of the volcano melting snow and ice.

land breeze   A breeze occurring at night when the sea is relatively warmer than the land and when, as a result, pressure is relatively lower over the sea. During the day opposite conditions prevail: the land is relatively warmer and the

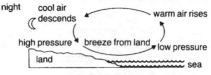

Land breeze Formation of a land breeze.

pressure gradient is from land to sea; a sea breeze then occurs. Such conditions arise because the land heats up and cools down more quickly than the sea.

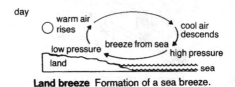

Land breeze Formation of a sea breeze.

landform Any natural feature of the earth's surface, such as mountains or valleys.

land reform The process of redistributing land, especially in developing nations where current circumstances may be against fair access to land and against agricultural improvement. For example, in parts of Latin America the traditional **land tenure** system is made up of large commercial estates (known as *estancias*) and small peasant holdings. Often the estate occupies the best farming land whilst the peasants farm in difficult conditions such as on valley sides. This inequality has led to pressure for land reform, sometimes by violent revolution. The aim of recent land-reform schemes has been to provide typical farming families with security, reasonable farming land, and an incentive to improve productivity. Concern for land reform occurs not only in situations of unequal distribution as described above, but also in traditional rural economies where land is allocated through the chieftainship, often in small and fragmented parcels, as occurs in parts of Africa. Here the necessary reforms would be consolidation of holdings, the provision of secure tenure, and some facility for the progressive farmer to improve productivity. However, clumsy land reform may do more harm than good: land in many rural societies is much more than a commodity, it is an integral part of the culture and heritage.

land tenure A system of land ownership or allocation.

land use The function of an area of land. For example, the land use in rural areas could be farming or forestry, whereas urban land use could be housing or industry. Sometimes land uses can come into conflict; for example, the quarrying of limestone in a **national park** would be in conflict with the aims of **conservation** and public access within the park.

land-value gradient The decline of average

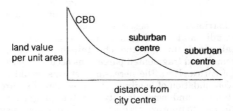

Land-value gradient The gradient for a typical city.

land values per unit area with increasing distance from the **CBD**. Whilst this is usually true, some suburban areas with good **accessibility** may prove to be good locations for business and retail centres. Therefore the land-value gradient may show peaks within suburban areas.

lapse rate The rate at which temperature changes with altitude. It is usual for temperature to decrease with altitude at an average rate of 0.6°C per 100 metres. In certain conditions a temperature inversion happens, with temperature increasing with altitude. Such an inversion can trap pollutants close to the surface of the Earth.

laterite A hard (literally 'brick-like') soil in tropical regions caused by the baking of the upper **horizons** by exposure to the sun. Laterite occurs often as a result of mismanagement of the tropical **environment**, for example by clear-felling of **tropical rain forest**, or by a decrease in the **regeneration** cycle in **shifting agriculture**.

Laterite has low agricultural potential; it is difficult to cultivate with simple tools and has a low nutrient content owing to the removal of forest and the consequent reduction in nutrient cycling.

latitude Distance north or south of the equator, as measured by degrees of the angle at the

Earth's centre:

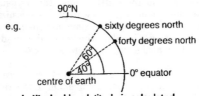

Latitude How latitude is calculated.

Laurasia The northern hemisphere supercontinent, consisting of the present North America, Europe and Asia (excluding India), that split from **Pangaea** *c* 200 million years ago. Laurasia is part of the theory of **continental drift**. See also **plate tectonics**.

lava **Magma** extruded onto the Earth's surface via some form of volcanic eruption. Lava varies in viscosity (see **viscous lava**), colour and chemical composition. Acidic lavas tend to be viscous and flow slowly, basic lavas tend to be nonviscous and flow quickly. Commonly, **lava flows** comprise basaltic material, as for example in the process of sea-floor spreading (see **plate tectonics**).

lava flow A stream of **lava** issuing from some form of volcanic eruption. See also **viscous lava**.

lava plateau A relatively flat upland composed of layer upon layer of approximately horizontally bedded lavas. An example of this is the Deccan Plateau of India.

leaching The process by which soluble substances such as mineral salts are washed out of the upper soil layer into the lower layer by rain water.

least-cost location The optimum location for an industry. The least-cost location may change over time with transport developments, changes in the sources of raw materials and changing markets. The least-cost location for iron and steel manufacture, for example, has shifted from inland coalfields to coastal sites as imported **raw materials** have increased in importance. See also **industrial location**.

levée The bank of a river, raised above the

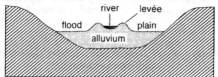

Levée Cross section of the lower stage of a river's course, showing height above flood plain.

general level of the **flood plain** by **sediment** deposition during flooding. When the river bursts its banks, relatively coarse sediment is deposited first, and recurrent flooding builds up the river's banks accordingly.

Note that continuous deposition on a river bed raises the entire channel above the general level of the flood plain.

light industry   Industry on a relatively small scale, as contrasted with **heavy industry**. Light industry includes industries such as the manufacture of electrical goods, printing and publishing, distribution trades and clothing manufacture.

Light industry is generally to be found on purpose-built estates, often on the edge of an urban area where congestion is less and **communications** are good. Light industry is generally cleaner and less polluting than heavy industry, and is organized mainly for the manufacture of consumer durables.

lignite   A soft form of **coal**, harder than **peat** but softer than **bituminous coal.**

limb   One element of a **fold**, with respect to either an **anticline** or a **syncline**.

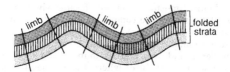

Limb Limbs in a section of folded strata.

limestone Calcium-rich **sedimentary rock** formed by the accumulation of the skeletal matter of marine organisms. The most favourable conditions for limestone formation comprise a warm, clear, shallow sea.

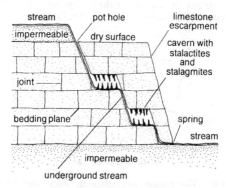

Limestone Typical features.

Many limestones contain fossils: shells and skeletal traces are common. Being both soluble and **permeable**, limestone gives rise to a dry surface landscape with characteristic features of underground **drainage**, as found in the Malham district of Yorkshire, where the Carboniferous limestone, approximately 350 million years old, is the major element of the landscape.

limestone pavement An exposed **limestone** surface on which the **joints** have been enlarged by the action of rainwater dissolving the limestone to form weak carbonic acid. These enlarged joints, or **grykes**, separate roughly rectangular blocks of limestone called **clints**.

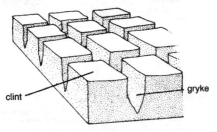

Limestone pavement

link The connection between two **nodes** in a communication **network**.

Little Ice Age A period from 1550 to 1850 characterized by extended cold seasons, heavy **precipitation** and the expansion of **glaciers**. It is thought that this may have occurred because of an absence of sunspot activity at this time. The reasons for this are yet to be explained.

load The **sediment** transported by the agents of **erosion** – rivers, moving ice, and the sea. The size and volume of load transported depends upon the power of the transporting medium. In a river system, for example, more load is carried in times of high **discharge**. A river's load comprises material rolled or bounced along the bed, material carried in suspension, and material carried in solution. The finest sediment is carried the greatest distances and may contribute to the formation of, for example, a **delta**.

loess A very fine **silt** deposit, often of considerable thickness, transported by the wind prior to **deposition**. In northern China, for example, there are extensive loess deposits which have been carried by wind from the arid plateau lands of Central Asia. Loess is extremely porous, and the surface is consequently dry. When irrigated, loess can be very fertile and consequently high **yields** can be obtained from crops grown on loess deposits.

location The position of population, settlement and economic activity in an area or areas. Location is a basic theme in **human geography**.

location triangle See **industrial location**.

longitude A measure of distance on the Earth's surface east or west of the Greenwich Meridian, an imaginary line running from pole to pole through Greenwich in London. Longitude, like **latitude**, is measured in degrees of an angle taken from the centre of the Earth. Lines of longitude are examples of *Great Circles*; a Great Circle is any circle drawn on the Earth's surface,

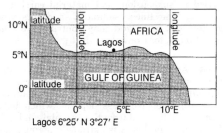

Lagos 6°25′ N 3°27′ E

(60′ [60 minutes] = 1°)

Longitude A grid showing the location of Lagos, Nigeria.

the centre of which coincides with the centre of the Earth. The Great Circle route between two points on the Earth's surface is the shortest possible route. The only line of latitude which is a Great Circle is the equator.

The precise location of a place can be given by a **grid reference** comprising longitude and latitude. See also **map projection**.

longshore drift The net movement of material along a beach due to the oblique approach of waves to the shore. Beach deposits move in a zig-zag fashion as shown in the diagram below. Longshore drift is especially active on long, straight coastlines. The movement of beach material can be halted by the construction of **breakwaters**.

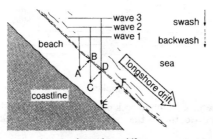

Longshore drift

As waves approach, sand is carried up the beach by the swash, and retreats back down the beach with the backwash. Swash occurs at 90° to the wave front; backwash (due purely to gravity) occurs directly down the slope of the beach. Thus a single representative grain of sand will migrate in the pattern A, B, C, D, E, F in the diagram. The resulting net movement of beach material is known as 'longshore drift'.

lopolith A basin-shaped igneous **intrusion**. Many lopoliths are small, though the Bushveldt lopolith of South Africa is exposed over an area of 65,000 km².

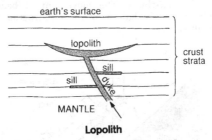

Lopolith

magma Molten rock originating in the Earth's **mantle**, that is the source of all **igneous rocks**.

malnutrition The condition of being poorly

nourished, as contrasted with **undernutrition**, which is lack of a sufficient quantity of food. The diet of a malnourished person may be high in starchy foods but is invariably low in protein and essential minerals and vitamins. Malnutrition occurs in many parts of Africa and Asia and gives rise to numerous ailments which lead to high infant mortality: kwashiorkor, marasmus, beriberi and pellagra are examples. The remedy for almost all such diseases is simple yet tragically unavailable to a large proportion of the world's population – high protein foods (such as fish, meat, pulses) and green vegetables to provide minerals and vitamins.

mantle The largest of the concentric zones of the Earth's structure, overlying the **core** and surrounded in turn by the **crust**.

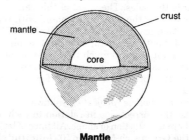

Mantle

manufacturing industry The making of articles using physical labour or machinery, especially on a large scale. See **secondary sector**.

map projection A method by which the curved surface of the Earth is shown on a flat surface map. As it is not possible to show all the Earth's features accurately on a flat surface, some projections aim to show direction accurately at the expense of area, some the shape of the land and oceans, while others show correct area at the expense of accurate shape.

One of the projections most commonly used is the *Mercator projection*, devised in 1569, in which all lines of **latitude** are the same length as the equator. This results in increased distortion of area, moving from the equator towards the poles. The effect of this can be seen by comparing the size of Greenland with that of South America. Greenland actually has 1/12 the area of South America, but on the Mercator projection it appears much the larger of the two. However, **compass** bearings are correct. This makes the projection suitable for navigation charts.

The *Mollweide projection* shows the land masses the correct size in relation to each other but there is distortion of shape. All lines of latitude and the Greenwich Meridian (0° **longitude**) are

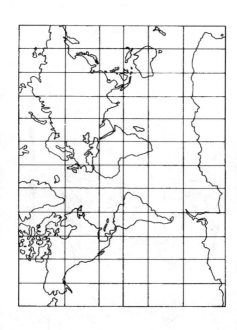

Map projection Mercator projection

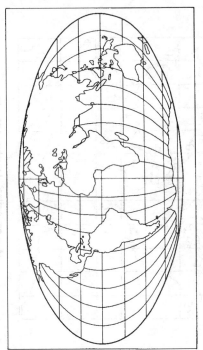

Map projection Mollweide projection

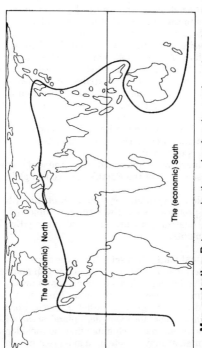

The (economic) North

The (economic) South

Map projection Peters projection, showing the north and south economic division of the world.

shown as straight lines, with the latter at right angles to the former. All other lines of longitude, and the land masses with them, are curved, with increasing curvature away from the Greenwich Meridian. As the Mollweide projection has no area distortion it is useful for showing distributions such as population distribution. However, because of its distortion of shape, direction and distance, it is not suitable for navigation charts.

The *Peters projection* is also an equal area projection, in which the Earth is divided into 100 rectangular fields. Direction is maintained but shape is distorted.

It must be emphasized that the only true representation of the Earth's surface is a globe.

marble A whitish, crystalline **metamorphic rock** produced when **limestone** is subjected to great heat or pressure (or both) during Earth movements.

marginal land In **agriculture** the term is used to describe land which is only just worth managing. The farmer will make only a small profit, if any, in the use of such land. Land may be regarded as marginal in either a physical or an economic sense: for example, upland areas with thin soils and harsh climates may be considered marginal, as may remote land at a great distance from (or inaccessible to) the market. Much of the rough upland pasture which is an element of British **hill farming** is marginal, and farmers stay in business with the aid of government subsidies.

maritime climate   A **temperate climate** that is affected by the closeness of the sea, giving a small annual range of temperatures – a coolish summer and a mild winter – and rainfall throughout the year. Britain has a maritime climate. Compare **extreme climate**.

market area   The area from which consumers will travel into a **central place** in order to obtain goods and **services**. The size of the market area is determined by the position of the central place in the **settlement hierarchy**: the market area (or **sphere of influence**) for low-order central places is small, since individual travel tolerance for everyday goods and services is low. Conversely, the market area for high-order central places is large, since individual travel tolerance for specialized goods and services is high. See **central place theory**.

market gardening   An intensive type of **agriculture** traditionally located on the margins of urban areas to supply fresh produce on a daily basis to the city population. Typical market-garden produce includes salad crops such as tomatoes, lettuce, cucumber, etc., cut flowers, fruit and some green vegetables. The classic market garden is characterized by a small landholding, high capital investment in the form of equipment such as greenhouses and fertilizers, plus high yields: these conditions are in response to the high value of land near urban areas and the perishability of the produce (see **Von Thünen theory**). However, modern developments in **communications** and transport technology, and indeed in food processing and consumer tastes, have led to a relaxation of the traditional locational factors behind market gardening. The production of, e.g. peas, beans and cauliflowers in the English fens, primarily for canning and freezing, is an example of the 'new' market gardening; some produce is marketed direct to London and to supermarkets via the excellent communications of eastern lowland England.

Although market gardening is still important in Great Britain, improved air communications have meant that much market-garden produce can be flown in from abroad, where it is grown more economically. For example, tomatoes are flown in from Spain, new potatoes from Egypt, and vegetables from Kenya (providing fresh food out of its normal season).

maximum and minimum thermometer An instrument for recording the highest and lowest

temperatures over a 24-hour period, the readings
usually being taken at 0800 hrs.

As the temperature rises, the alcohol in the
left-hand tube expands, pushing the mercury up
in the right-hand tube, in which the alcohol also
heats up and vaporizes into the cavity in the
conical bulb. When the temperature falls this
vapour liquefies and the alcohol in the left-hand
tube contracts, causing the mercury to flow in
the reverse direction. Metal pointers mark the
extreme positions of the mercury column giving
the maximum temperature on the right-hand
column and the minimum on the left; these are
reset with a magnet.

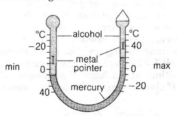

Maximum and minimum thermometer

meander A large bend, especially in the middle
or lower stages of a river's course. See **flood
plain**. A meander is the result of lateral
corrasion which becomes dominant over

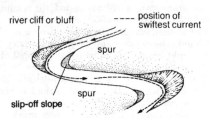

Meander Features of a river meander.

vertical corrasion as the **gradient** of the river's course decreases. The characteristic features of a meander are summarized thus:

Concentration of **erosion** on the outer bank causes undercutting and the development of a **river cliff**. Slack water on the inner bank causes **deposition**, and a bank of **sediment** called a **slip-off slope** is produced. See also **oxbow lake**.

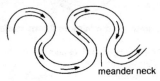

Meander Fully formed meanders.

mesa A flat-topped, isolated hill in arid regions, for example in Arizona and New Mexico in the USA. A mesa has a protective cap of hard **rock** (that resists erosion by wind) underlain by softer, more readily eroded **sedimentary rock**. A butte is a relatively small outlier of a mesa. Unprotected soft rock is eroded, leaving the mesas and buttes upstanding.

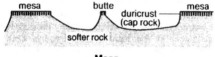

Mesa

metamorphic rock A **rock** which has been changed by intensive heat or pressure. Metamorphism implies an increase in hardness and resistance to **erosion**. Shale, for example, may be metamorphosed by pressure into **slate**; sandstone by heat into **quartzite, limestone** into **marble**. Metamorphism of pre-existing rocks is associated with the processes of **folding, faulting** and **vulcanicity**.

migration A permanent or semipermanent change of residence. **Population migration** may occur at a variety of scales and for a number of reasons: international, regional and local migrations occur, and their precise reason for movement will differ between individuals. Employment-regulated migration, which occurs in the current **Third World** *urbanization process*, is dominant, but people also move for family reasons, for retirement, or as a result of political pressure (forced migration), as has happened in Bosnia and Kosovo. Scale contrasts are considerable: at the international scale, movement from the British Isles to North America in the 19th century involved many thousands of individuals; at the local scale there has been a tendency for people to move from the inner to the outer city, and even out of the city altogether. The decision to migrate, whatever the scale, is determined by the balance of **push factors, pull factors** and *intervening obstacles*. Push factors are negative aspects of life at the current place of residence; pull factors are attractive aspects of some potential future location; intervening obstacles might include such factors as distance, cost, physical and political barriers – the greater the magnitude of the obstacle, the less likely the migration will be. Invariably there is *distance-decay* in migration, i.e. the number of migrants into a given destination declines with the increasing distance of origins.

mixed farming   A system of **agriculture** which comprises a variety of arable and pastoral elements.

monoculture The growing of a single crop. Traditional grape cultivation for wine in the south of France is an example. Monoculture can be unsound in two ways: firstly the **soil** is progressively drained of specific nutrients, and secondly dependence on a single crop may be dangerous in the market if the crop fails or if consumer tastes change. Monoculture may also cause an increase in plant-specific pests and diseases.

monsoon The term strictly means 'seasonal wind' and is used generally to describe a situation where there is a reversal of wind direction from one season to another. This is especially the case in South and Southeast Asia, where two monsoon winds occur, both related to the extreme pressure gradients created by the large land mass of the Asian continent. In summer (April to September), the intense heating of the land leads to the development of low pressure over northwestern India, and southwesterly winds are drawn in over the Indian ocean. This southwest monsoon brings heavy rain to India and Southwest Asia in general, and especially to those areas where the **orographic** effect is operating. In winter (October to March), the chilling of the Asian interior leads to high pressure and the establishment of the northeast monsoon: cold dry winds which bring little rain. In northern China they may also carry dust from the deserts of the continental interior.

moraine A collective term for debris deposited on or by **glaciers** and ice bodies in general. Moraine differs from fluvial **sediment** in being unsorted and composed of angular fragments. Several types of moraine are recognized: *lateral* moraine forms along the edges of a valley glacier where debris eroded from the valley sides, or weathered from the slopes above the glacier, collects; *medial* moraine forms where two lateral moraines meet at a glacier junction; *englacial* moraine is material which is trapped within the body of the glacier; and *ground* moraine is material eroded from the floor of the valley and used by the glacier as an abrasive tool. A **terminal** moraine is material bulldozed by the

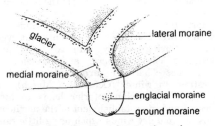

Moraine Position of types of moraine.

glacier during its advance and deposited at its maximum down-valley extent. *Recessional* moraines may be deposited at standstills during a period of general glacial retreat.

mortlake See **oxbow lake**.

multiple nuclei model A model of urban structure devised by Harris and Ullman in 1945, stating that most large cities develop around a

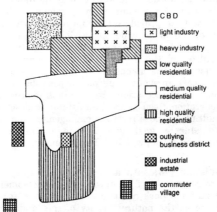

CBD

light industry

heavy industry

low quality residential

medium quality residential

high quality residential

outlying business district

industrial estate

commuter village

Multiple nuclei model

number of separate centres or nuclei, rather than round a single centre (compare **Burgess model** and **sector model**). Different land uses are therefore situated around the city, creating a cellular structure. The pattern of these cells or nuclei reflects the unique factors of the site and/or history of any particular city. See also **shanty town**.

multiple-purpose resource management A strategy for resource management which attempts to provide maximum availability of a **resource** to as wide as possible a variety of users, without endangering the quantity or quality of the resource for any particular consumer. The management of water resources is a case in point: multiple-purpose dams can cater for recreation, **hydroelectric power** generation, flood control and fishing. In practice, multiple-purpose resource management may be an elusive goal due to the problems of accommodating widely different demands.

national park An area of scenic countryside protected by law from uncontrolled development. A national park has two main functions: (a) to conserve the natural beauty of the landscape; (b) to enable the public to visit and enjoy the countryside for leisure and recreation.

natural arch   A feature of coastal **erosion** resulting from the differential erosion of hard and soft **rocks**. For example, softer rocks in a **headland** may be rapidly eroded, first to produce two caves and later to create a natural arch as the backs of the two caves are eroded through. Eventually the roof of the arch may collapse to produce a **stack**.

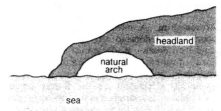

Natural arch

natural hazard   A natural event which, in extreme cases, can lead to loss of life and destruc-tion of property. Some natural hazards result from geological events, such as **earthquakes** and the eruption of **volcanoes**, whilst others are due to weather events such as **hurricanes**, floods and droughts. Natural hazards often have their greatest impact in developing countries which are least able to cope with the aftermath of such events.

natural increase   The increase in population due to the difference between **birth rate** and **death rate**: in the **demographic transition**, for example, the phase of rapid population increase is the result of a constant high birth rate and a rapidly falling death rate. Natural increase is one component of *gross* population increase: in-migration can also cause a rise in total population. The complete **population change** equation for a nation or region will thus comprise both natural and migrational factors.

neap tides   See **tides**.

nearest neighbour analysis   See **spatial distribution**.

neighbourhood amenities   Useful facilities in the local area. Many neighbourhood amenities are provided by the local council, e.g. swimming pools, park benches, bus shelters and street lights. See also **household amenities**.

neocolonialism   A maintenance of the authority/dependency relationship of the colonial period through economic mechanisms. Although the majority of former colonies, of Britain for example, have gained political independence, they remain economically dependent on the former colonial power. Many industrial nations maintain close connections with their former colonies, often through **transnational corporations** and trading links. Products such as tea, coffee, tin and copper have their prices fixed in the industrial nations and not in the producing nations. Thus, although the period of political colonialism is largely over, economic colonialism persists.

network A system of **links** and **nodes** via which flows of communication are passed. The structure of networks is represented diagrammatically as follows:

See also **accessibility matrix, Beta index, connectivity**.

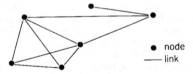

Network Nodes and links.

nevé Compact snow. In a **corrie** icefield, for example, four layers are recognized: blue and white ice at the bottom of the ice mass; nevé overlying the ice, and powder snow on the surface.

newly industrialized country (NIC) A **developing country** which is becoming industrialized, for example Malaysia and Thailand. Some NICs have successfully used large-scale development to move into the industrialized world. Usually the capital for such developments comes from outside the country. Some NICs, such as South Korea and Taiwan, have now moved on to high-technology industries, though again finance (and expertise) has been brought in from outside.

new town A new urban location created (a) to provide overspill accommodation for a large city or **conurbation**; (b) to provide a new focus for industrial development, for example in a **depressed region**.

In the UK, 26 new towns were created between 1945 and 1971, designed to take people from overcrowded areas of large cities and place them in new communities with all facilities provided. Planned new towns included Stevenage, Harlow, Basildon and Milton Keynes to take population from London and Telford, Skelmersdale and Glenrothes created to provide foci for **regional development**. There are now no new towns under construction and government investment is being redirected to decaying inner city areas.

nivation A type of physical **weathering**

whereby **rocks** are denuded by the freezing of water in cracks and crevices on the rock face. Water expands on freezing into ice, and this process causes stress and fracture along any line of weakness in the rock. Nivation debris accumulates at the bottom of a rock face as **scree**. Nivation is particularly active in cold **climates** or at altitude, for example, on the exposed slopes above a **glacier**.

nodal region   An area defined by communication links radiating from a **node**. For example, in the nodal region of the Paris Basin all communication links spread out from the French capital city, Paris, which thus dominates the region.

node   A **central place**; in **network** studies **settlements** are referred to as nodes.

nomadic pastoralism   A system of **agriculture** in dry grassland regions; examples of societies which are based on nomadic pastoralism include the Masai of East Africa and the Fulani of northern Nigeria. People and stock (cattle, sheep, goats) are continually moving in search of pasture and water: the low rainfall of much of the African **savannas** and the scanty nature of the grazing vegetation require a nomadic cycle of activity. The pastoralists subsist on meat, milk and other animal products.

Nomadic pastoralism is under pressure from a variety of causes, for example severe drought in the Sahel region (the southern margins of the Sahara desert) in the late 1970s led to destitution among thousands of wandering herdspeople. Overgrazing and the trampling of the soil surface around boreholes has hastened the process of **desertification**.

Various schemes sponsored, for example, by the United Nations, have been introduced in an attempt to improve conditions for the pastoralists, for example the introduction of high-quality stock, parasite control and pasture improvement. The chief problem is that many of these schemes require a fundamental change in the nomad's traditional lifestyle: permanent **settlement** is an integral part of most development schemes.

non-governmental organizations (NGOs) Independent organizations, such as charities (Oxfam, Water Aid) that provide aid and expertise to economically developing countries.

nonrenewable resources Resources of which there is a fixed supply, which will eventually be exhausted. Examples of these are metal ores and **fossil fuels**. Compare **renewable resources**.

north and south A way of dividing the industrialized nations, found predominantly in the north – Europe, North America and Japan, plus Australia – from those less developed nations in the south – South America, Africa and parts of Asia. The term 'north and south' is taken from the Brandt Report on the inequalities of wealth between nations, published in 1980, which, by using a number of indicators, distinguished the richer countries of the world from the poorer countries. The gap which exists between the rich 'north' and the poor 'south' is called the *development gap*. See figure in **map projections** (Peters projection map).

northings   An element of a **grid reference**. See **eastings**.

nuclear power station   An electricity-generating plant using nuclear fuel as an alternative to the conventional **fossil fuels** of **coal**, oil and gas. Nuclear power stations, while expensive to construct, are relatively cheap to run but have to be built in remote (often coastal) locations well away from population concentrations. This is partly in response to public anxiety over the safety of such stations and partly due to the problems of radioactive waste disposal. The fuel used in most nuclear power stations is the element uranium, world **reserves** of which are extensive.

nuée ardente A very hot and fast-moving cloud of gas, ash and rock that flows close to the ground after a violent ejection from a volcano. It is very destructive.

nunatak A mountain peak projecting above the general level of the ice near the edge of an **ice sheet**. Such features occur, for example, in Greenland and Antarctica.

nutrient cycle The cycling of nutrients through the **environment** as, for example, in the process of leaf-fall in a woodland **ecosystem**. Here, fallen vegetation is broken down by bacteria into constituent nutrients such as calcium and magnesium; these are released into the **soil** where they are available for root uptake and renewed growth of vegetation, and so the cycle continues on an annual basis. The actual pathways of nutrients through an ecosystem are very complex, but all involve the use of nutrients by living organisms for growth, and the subsequent release of nutrients back to the environment on death and decay.

ocean current A movement of the surface water of an ocean. The most important causes of this movement are the **prevailing winds** and the differences in water density due to temperature or salinity. The shape of the continents and the rotation of the Earth can also influence the direction of currents. Ocean currents caused by prevailing winds are called drift currents, the best known being the Gulf stream. Between the equator and the temperate regions in the northern hemisphere, the circulation of ocean currents is clockwise, whereas in the southern hemisphere it is anticlockwise.

offshore bar A low bank of sand and shingle lying some distance offshore and exposed at high tide. The offshore bar is created when a very gently shelving seabed causes waves to break well away from the actual shoreline. The Cape Hatteras coastline of the Atlantic coast of the USA is an area where such conditions prevail (See diagram overleaf).

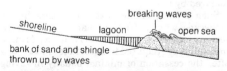

Offshore bar The formation of an offshore bar.

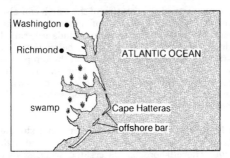

Offshore bar The Cape Hatteras coastline.

opencast mining   A type of mining where the mineral is extracted by direct excavation rather than by shaft or drift methods. For example, in parts of the Yorkshire coalfield the coal measures occur very close to the surface and the superficial overburden is relatively easily removed. The **coal** is then excavated by mechanical grabs and removed by trucks.

Such mining creates extensive scars in the landscape which, if left unmanaged, represent serious environmental deterioration. Mining companies are required to undertake *landscaping* after the cessation of mining, usually involving the infilling of the opencast site and the planting of vegetation on the reclaimed surface.

The vegetation also helps to stabilize the infill.

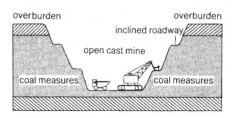

overburden

overburden

inclined roadway

open cast mine

coal measures

coal measures

Opencast mining

Ordnance Survey map   A map produced by the Ordnance Survey, the government agency responsible for surveying and mapping the United Kingdom. The Ordnance Survey also publishes a range of maps for the general public, at a variety of scales, that give information for particular purposes such as motoring, walking and sightseeing. The two most commonly used OS maps are the 'Pathfinder' 1:25 000 map and the 'Landranger' 1:50 000 map.

organic fraction   That proportion of the **soil** which is composed of material derived from the breakdown of vegetation or other organic matter. Compare **inorganic fraction**. See also **horizon**.

organic farming A system of farming that avoids the use of any **artificial fertilizers** or chemical pesticides, using only **organic fertilizers** and pesticides derived directly from animal or vegetable matter. Yields from organic farming are lower, but the products are sold at a premium price.

organic fertilizer A fertilizer composed of organic material, e.g. horse manure, farmyard manure, seaweed derivatives and bonemeal. Compare **artificial fertilizer**.

orogeny A geological period of **fold mountain**

Orogeny	Peak date	Examples
Alpine or Tertiary	Approx 50 million years ago	Rockies, North America Andes, South America Alps, Europe Himalayas, Asia
Hercynian	Approx 300 million years ago	Appalachians, USA
Caledonian	Approx 400 million years ago	Scottish Highlands

Orogeny Table showing three important orogenies.

building activity. There are three generally recognized orogenies.

orographic rainfall or **relief rainfall** Precipitation caused by the rising of air over, for example, a coastal mountain range.

The area inland from the coastal range is likely to experience very low rainfall since the air has lost its moisture on the seaward flank of the mountains. This is called a *rain shadow* area. Such a situation occurs on the western seaboard of North America where the Rocky Mountains generate the orographic effect.

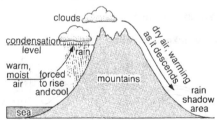

Orographic rainfall

outwash Sedimentary material deposited by meltwater issuing from a **glacier** snout or **ice sheet** margin.

Outwash sands and gravels differ from glacial

moraines in being *sorted* and subjected to **attrition**. The coarsest **sediments** are deposited close to the ice margin; fine material will be transported greater distances before **deposition**.

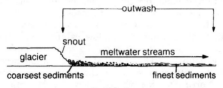

Outwash

overfold   See **fold**.

oxbow lake, mortlake or **cut-off**   A crescent-

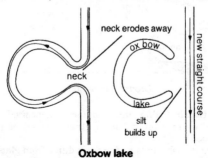

Oxbow lake

shaped lake originating in a **meander** that was abandoned when **erosion** breached the neck between bends, allowing the stream to flow straight on, bypassing the meander. The ends of the meander rapidly silt up and it becomes separated from the river. In time the entire oxbow lake will silt up, pioneer vegetation will invade, and the lake will eventually disappear.

ozone A form of oxygen found in a layer in the **stratosphere**, where it protects the Earth's surface from ultraviolet rays. There has been recent concern about holes appearing in the ozone layer over the polar regions. There are many possible reasons why these holes might occur, but it is known that the ozone layer is damaged by the chlorine released by **CFCs**. An increase in the ultraviolet rays reaching the Earth could result in an increase in skin cancers, and damage to crops, vegetation, livestock and wildlife in countries near the polar regions.

Pangaea The supercontinent or universal land mass in which all continents were joined together approximately 200 million years ago. The theory of Pangaea's existence was devised by Alfred Wegener in 1912. See **continental drift**.

pastoral farming A system of farming in which the raising of livestock is the dominant element.

In the commercial context this may be, for example, dairy farming in Britain or sheep rearing in Australia, while subsistence pastoralism occurs in many parts of the **Third World**. See also **nomadic pastoralism**.

peasant agriculture   The growing of crops or raising of animals, partly for subsistence needs and partly for market sale. Peasant agriculture is thus an intermediate stage between subsistence and commercial farming.

peat   Partially decayed and compressed vegetative matter accumulating in areas of high rainfall and/or poor **drainage**.

In Britain, peat occurs in the upland areas of the north and west, forming *blanket-bog* over much of the Pennines, and in low-lying parts of East Anglia. Peat **soil** results from the limited break-down of fallen vegetation in *anaerobic* conditions: total breakdown into humus cannot occur in waterlogged, airless conditions. Upland peat is acidic and infertile due to **leaching**; lowland peat has a higher nutrient status and is more useful for **agriculture**.

The removal of lowland peat for use as a garden fertilizer is currently a controversial issue. The large-scale removal of peat disrupts the natural ecosystem, and **conservation** measures are now being looked at to preserve many of Britain's peat lands.

pedestal rock   See **Zeugen**.

peneplain   A region that has been eroded until it is almost level. The more resistant rocks will stand above the general level of the land.

per capita income   The GNP (gross national product or national income) of a country divided by the size of its population. It gives the average income per head of the population if the national income were shared out equally. Per capita income comparisons are used as one indicator of levels of economic development.

percolation   The movement of water through soil pores and rock crevices.

periglacial features   A periglacial landscape is one which has not been glaciated *per se*, but which is affected by the severe **climate** prevailing around the ice margin. Intensive **nivation** is characteristic of periglacial **environments**, as is *solifluction*, a process whereby thawed surface soil creeps downslope over a permanently frozen **subsoil** (*permafrost*). Much of the Canadian **tundra** and Siberian heartland are affected by such periglaciation.

periphery   A remote and/or underprivileged region as in the core/periphery model (see **core**).

Such regions are generally lacking in resources and offer little development opportunity, and as such are the last to be integrated into the national development process.

permafrost   The permanently frozen subsoil that is a feature of areas of Tundra.

permeable rock   Rock through which water can pass via a network of pores between the grains. Examples are sandstone and chalk. Compare **pervious rock**. See also **impermeable rock**.

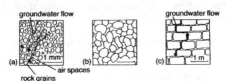

permeable rock (a) Permeable rock. (b) Impermeable rock. (c) Pervious rock.

pervious rock   Rock which, even if nonporous, can allow water to pass through via interconnected **joints, bedding planes** and **fissures**. An example is **limestone**. Compare

permeable rock. See also **impervious rock**.

pH A measure of acidity/alkalinity. A pH value of 7.0 is regarded as neutral, while pH values of less than 7.0 indicate acidic conditions. pH values greater than 7.0 indicate increasingly alkaline conditions. pH tests are used to assess the acidity of soil and of water. See **acid rain**. The optimum soil pH for cereal growth is about 6.5.

photosynthesis The process by which green plants make carbohydrates from carbon dioxide and water, and give off oxygen. Photosynthesis balances **respiration**. However, the burning of **fossil fuels** has greatly increased the amount of carbon dioxide in the atmosphere, and **deforestation** has seriously reduced the number of trees available to release oxygen, so that the natural balance has been lost.
See **global warming**.

physical geography The study of our **environment**, comprising such elements as geomorphology, hydrology, pedology, meteorology, climatology and biogeography.

pie chart A circular graph for displaying values as proportions (see over):

Journeys to work
(sample of urban population)

Mode	No.	%	Sector° (% × 3.6)
Foot	25	3.2	11.5
Cycle	10	1.3	4.7
Bus	86	11.2	40.3
Train	123	15.9	57.2
Car	530	68.5	246.3
Total	774	100	360
		percent	degrees

Pie chart A pie chart showing how a sample of urban population travels to work.

plantation   agriculture A   system   of
agriculture located in a tropical or semitropical
environment, producing commodities for
export to Europe, North America and other
industrialized regions. Coffee, tea, bananas,
rubber and sisal are examples of plantation
crops.

Plantation agriculture is distinctive in that it
is a form of **commercial agriculture** located in
a generally subsistence or peasant environment:
it is an extension of the commercial agriculture
of the developed world into a mainly **Third
World** environment. Some plantations are run
and financed by **transnational corporations**
and the profits from such operations are
generally channelled back to Europe or North

America. As a result many plantations are institutions of **neocolonialism**. There is a worry that plantations often take up valuable farmland, growing commodities required by the richer countries. Thus more valuable local crops are forced onto poorer land. In areas where unemployment is high, plantations have traditionally paid low wages. On a more positive side, many plantation operators provide such facilities as housing, education and health care for their workers, as well as a plot of land. But it is difficult to avoid the conclusion that plantation agriculture in its traditional form is unacceptable in view of contemporary development priorities in Third World nations.

plate tectonics The theory that the Earth's **crust** is divided into seven large, rigid plates, and several smaller ones, that are moving relative to each other over the upper layers of the Earth's **mantle**. See **continental drift**. **Earthquakes** and volcanic activity occur at the boundaries between the plates. See page 302.

There are three types of plate boundary:
(a) *constructive plate boundary* – where new oceanic crust is formed on either side of the boundary as **magma** wells up (see **convection**), eventually increasing the width of the ocean bed by a few centimetres per year (*sea-floor spreading*) and thus helping to push the continents on

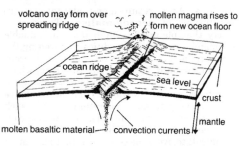

volcano may form over spreading ridge
molten magma rises to form new ocean floor
ocean ridge
sea level
crust
mantle
molten basaltic material
convection currents

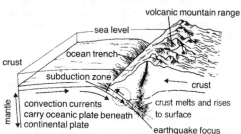

volcanic mountain range
sea level
crust
ocean trench
subduction zone
crust
mantle
convection currents carry oceanic plate beneath continental plate
crust melts and rises to surface
earthquake focus

Plate tectonics (a) Constructive plate boundary. (b) Destructive plate boundary.

either side further apart.
(b) *destructive plate boundary* – where the oceanic crust is gradually carried down into the

mantle (*subduction*), producing great stresses that cause major earthquakes. Volcanic eruptions occur at the surface as molten rock rises up through the overlying plate. A destructive plate boundary is marked by an *oceanic trench*; the trenches in the Pacific, about 10 000 m deep, are the deepest places on Earth. Where the oceanic plate is subducted beneath a continental plate, the edge of the continental plate becomes compressed and folded. See **fold mountains**.
(c) *conservative plate boundary* – where crustal material is being neither created nor destroyed. Instead the plates slide past each other.

plucking A process of glacial **erosion** whereby, during the passage of a valley **glacier** or other ice body, ice forming in cracks and fissures drags out material from a **rock** face. This is particularly the case with the backwall of a **corrie**.

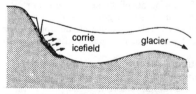

Plucking Blocks of rock are dragged out as ice tears away from a corrie wall.

plug The solidified material which seals the vent of a **volcano** after an eruption. A volcanic plug is thus responsible for the build-up of pressure which may result in an explosive eruption at some later stage. **Viscous lavas** produce the most effective plugs and, because of their resistance to **erosion**, volcanic plugs tend to stand out in the landscape when softer surrounding material has been worn away. A well known example is Edinburgh Castle Rock.

plunge pool See **waterfall**.

plutonic rock **Igneous rock** formed at depth in the Earth's **crust**; its crystals are large due to the slow rate of cooling. **Granite**, such as is found in **batholiths** and other deep-seated intrusions, is a common example.

podzol The characteristic **soil** of the **taiga** coniferous forests of Canada and northern Russia. Podzols are leached, greyish soils: iron and lime especially are leached out of the upper horizons, to be deposited as *hardpan* in the B **horizon**.

pollution Environmental damage caused by improper management of **resources**, or by careless human activity.

In the countries of the 'first world' (see **Third World**) progress has been made in the control of some of the worst causes of chemical pollution, such as persistent agricultural chemicals and heavy metals. Smoke control legislation has led to cleaner air in cities (see **smog**). However, emissions of sulphur dioxide from the burning of **fossil fuels** by industry, and of nitrous oxides from vehicle exhausts, remain high as governments and industries are reluctant to spend the large amounts of money necessary to reduce them. These pollutants are then blown elsewhere by the **prevailing winds**. Emissions from Britain, for example, are blown over Scandinavia, where the resulting **acid rain** has led to the poisoning of entire **ecosystems**: many lakes are now devoid of fish; forest growth is stunted; and public water supplies require calcification.

In the formerly communist countries of eastern Europe and the USSR, no controls were exerted over chemical pollution by industry or by vehicle exhaust emissions because production was the priority of the governments concerned. Their financial resources were invested in outdated industrial plant that produce enormous quantities of highly polluting emissions. Damage to plants, fish and animals was very serious in some areas, and the incidence of respiratory diseases in the human population of the industrial towns was distressingly high. It will take many years and huge amounts of money to reduce pollution levels, and meanwhile the damage to people and to the environment will continue.

In many **developing countries**, also, production is seen as the priority and the reduction of pollution as a luxury that the countries cannot afford.

Besides chemical and radioactive pollution (caused by nuclear accidents, as happened at the Chernobyl nuclear power station in the Ukraine in 1986), which may be life-threatening, there are 'nuisance' pollutions such as *noise pollution* by low-flying aircraft and the *visual pollution* of quarries, rubbish dumps, etc.

population change The increase of a population, the components of which are summarized in the diagram.

population density The number of people per unit area. Population densities are usually expressed per square kilometre, and can range from less than one in remote, inhospitable regions, to many hundreds in urban areas or on highly productive agricultural land.

population distribution The pattern of population location at a given **scale**. At the global scale, population distribution shows a concentration in specific areas, for example in parts of Asia and Europe, and is sparse in others, such as in the polar regions and the hot deserts.

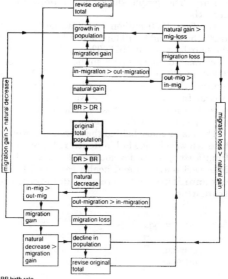

Population change Flow diagram for the population change system.

population explosion   On a global **scale**, the

dramatic increase in population during the 20th century. The graph shows world **population growth**.

The first 1000 million was reached by about 1820; the second by 1930; the third by 1960; and the fourth by about 1975. The world population is estimated to be 6,168 million in 2000 and 9,800 in 2050 (according to United Nations figures). The greatest growth is expected in Africa. In 1950 8.9% of the world's population was in Africa. In 2050 it is estimated that 21.8% of the population will live there.

The major cause behind the increasing growth rate is a world-wide fall in **death rate**, resulting from a complex of development factors such as improved nutrition and health care, and better **communications**.   See   **demographic transition**.

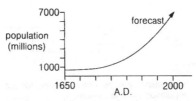

Population explosion Graph showing world population growth.

population growth An increase in the population of a given region. This may be the result of natural increase (more births than deaths) or of in-migration, or both.

population migration See **migration**.

population pyramid A type of **bar graph** used to show population structure, i.e. the age and sex composition of the population for a given region or nation. See diagrams below and overleaf.

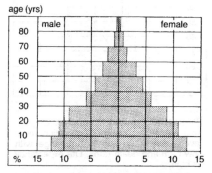

Population pyramid Pyramid for India, showing high **birth rates** and high **death rates**. Life expectancy is low and a large percentage of the population is young.

A population pyramid has data for males plotted on one side and data for females on the other.

The shape of a population pyramid is a useful indicator of the stage of development reached by a nation (see **demographic transition**) and can also indicate government policy on population planning and the impact of major events such as wars.

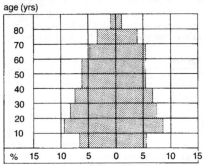

age (yrs)

Population pyramid Pyramid for England and Wales, showing a population more evenly distributed over the age range, with low birth and death rates. Life expectancy is high.

postindustrial Characteristic of an economy that is no longer based on **heavy industry** but is increasingly dominated by microelectronics, automation, and the service and information sectors. See **tertiary sector, quaternary sector**.

pothole   1. A deep hole in limestone, caused by the enlargement of a **joint** through the dissolving effect of rainwater. The Yorkshire Pennines contain classic examples of potholes, such as Gaping Gill near Ingleton, Yorkshire. Surface streams may disappear down certain potholes, which are then known as *sinkholes* or *swallow holes*, to become underground watercourses. **2.** A hollow scoured in a river bed by the swirling of pebbles and small boulders in eddies.

precipitation   Water deposited on the Earth's surface in the form of e.g. rain, snow, sleet, hail and dew. See **orographic rainfall**.

preindustrial   A term used to describe the early stages of the **development process**. Largely agricultural economies in which the foundations for development are being established, such as agricultural extension projects and improved **communications**, would be described as preindustrial. The implication of such terminology is that **industrialization** can be equated with development; while this has been true in recent history, it may be that alternative models will emerge as **resource** shortage and **pollution** become recognized globally.

prevailing wind The dominant wind direction of a region. In northwest Europe, for example, southwesterly winds predominate, i.e. occur more frequently than those from any other direction. Prevailing winds are named by the direction from which they blow.

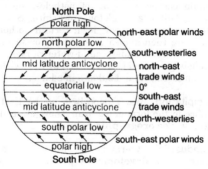

Prevailing wind

primary sector That sector of the national economy which deals with the production of primary materials: **agriculture**, mining, forestry and fishing. Primary products such as these have had no processing or manufacturing involvement. The total economy comprises the primary sector, the **secondary sector**, the

tertiary sector and the **quaternary sector**. The distribution of employment between these four sectors is a measure of the state of development of the nation: the early stages of the development process are marked by a concentration of the labour force in the primary sector, especially in agriculture, whilst the progress of development is characterized by increasing employment in, firstly, the secondary, and, later, the tertiary and quaternary sectors. It is also possible to observe regional variations within a country, e.g. the southeast of Britain contains a greater proportion of tertiary and quaternary workers than certain northern areas of Britain. A mining area contains a large number of primary workers, and an industrial region contains a large number of secondary workers.

Many **developing countries** are characterized by the employment of a large percentage of their working population in the primary sector, especially in **agriculture**. This is due partly to a lack of investment in manufacturing – many primary products are exported for processing to the developed nations. For example, a cash crop such as cocoa from Ghana is exported in its raw state to developed countries where it is manufactured into chocolate. The profits made by a developed nation on the sale of the finished chocolate are much greater than those made by the developing country on the raw cocoa.

primary source   See **secondary source**.

pull factors   See **migration**.

pumped storage   Water pumped back up to the storage lake of a **hydroelectric power** station, using surplus 'off-peak' electricity. The water can then be used again for power generation.

push factors   See **migration**.

pyramidal peak   A pointed mountain summit resulting from the headward extension of **corries** and **arêtes.** Under glacial conditions a given summit may develop corries on all sides, especially those facing north and east. As these erode into the summit, a formerly rounded profile may be changed into a pointed, steep-sided peak. The Matterhorn in the Alps is a classic example of this.

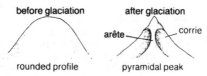

Pyramidal peak

pyroclasts   Rocky debris emitted during a volcanic eruption, usually following a previous emission of gases and prior to the outpouring of **lava** – although many eruptions do not reach the final lava stage. Pyroclasts may comprise lumps of solidified lava from previous eruptions, as found in a volcanic **plug** or chunks of *country rock*, i.e. the crustal material close to the volcanic vent, or finer debris such as ash and dust. The largest pyroclasts are *volcanic bombs*, pieces of debris that may weigh up to several tonnes. Pebble-sized debris is referred to as *lapilli*.

quality of life The level of wellbeing of a community and of the area in which the community lives.

quartz One of the commonest minerals found in the Earth's **crust**, and a form of silica (silicon+oxygen). Most **sandstones** are composed predominantly of quartz.

quartzite A very hard and resistant **rock** formed by the metamorphism of **sandstone**.

quaternary sector That sector of the economy providing information and expertise. This includes the microchip and microelectronics industries. Highly developed economies are seeing an increasing number of their workforce employed in this sector. Compare **primary sector, secondary sector, tertiary sector**.

rainfall interception The process by which plants and trees break the force of precipitation. Rain lands on the surfaces of the vegetation and fills the hollows. It then runs down the trunks and stems and drips from the leaves on to the ground, where it can sink in gently, instead of running off the surface.

The density and type of vegetation in an area are important in determining the speed at which rainwater moves through the landscape. Different types of vegetation intercept water to a greater or lesser degree. Coniferous forest, for example, intercepts 58% of the rain falling upon it; tall grasses intercept about 27%.

Where large tracts of natural vegetation have been removed from an area, splash erosion and sheet erosion occur (see **soil erosion**). Where soil on a slope is not consolidated by a network of roots, it can be loosened by rain and slip off, often with tremendous force, as a landslide. This has been a big problem in the Himalayas and areas of tropical rain forest, due to **deforestation**.

In urban areas, the lack of vegetation is compensated for by an artificial drainage system.

rain gauge An instrument used to measure rainfall. Rain falls through the funnel into the jar below and is then transferred to a measuring cylinder. The reading is in millimetres and indicates the depth of rain which has fallen over an area. Rain gauges are normally read once or twice a day.

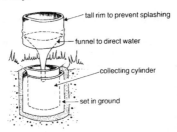

tall rim to prevent splashing

funnel to direct water

collecting cylinder

set in ground

Rain gauge

rain shadow   See **orographic rain**.

raised beach   See **wave-cut platform**.

range   The maximum distance a consumer is prepared to travel in order to purchase a good or service in a central place (see **central place theory**). The range of low-order, everyday goods such as bread, newspapers and daily groceries is very short. Consequently, journeys for these goods are frequent and people will only travel short distances for them. On the other hand, the range of high-order, specialized goods is much greater and people will therefore travel longer distances to purchase these items, and their journeys will be less frequent.

rapid   An area of broken, turbulent water in a river channel, caused by a stratum of resistant **rock** that dips downstream (see diagram). The softer rock immediately upstream and downstream erodes more quickly, leaving the resistant rock sticking up and creating an obstacle to the flow of the water. Compare **waterfall**.

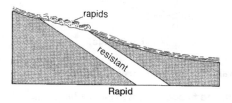

Rapid

raw materials   The **resources** supplied to industries for subsequent manufacturing processes, for example agricultural products, minerals and timber are **raw materials**. Many **primary products** are used as raw materials.

regeneration   Renewed growth of, for example, forest after felling. Forest regeneration is crucial to the long-term stability of many **resource** systems, from **bush fallowing** to commercial forestry.

region An area of land which has marked boundaries or unifying internal characteristics. Geographers may identify regions according to physical, climatic, political, economic or other factors. The Sahara desert is an example of a climatic region, while southeast England (London and the Home Counties) is an economic region.

regional development A planning policy made by a country to overcome regional differences in income, wealth, education, medical facilities, transport, etc. Such divisions between region must be minimized if countries are to achieve balanced development. Regional development plans can take many forms; for example, improvements in infrastructure (i.e. the construction of facilities that lay the foundations for further agricultural or industrial development – new roads, power supplies, etc.), resettlement schemes (such as the Indonesian transmigration scheme), or resource development (e.g. mineral exploitation).

Regional inequalities tend to be much more pronounced in **developing countries**, therefore such plans tend to be on a larger and more complex scale than those of developed countries.

rejuvenation Renewed vertical **corrasion** by rivers in their middle and lower courses, caused by a fall in sea level, or a rise in the level of land relative to the sea.

The point at which downcutting recommences (*knickpoint*) may be marked by a **waterfall**. **Meanders** will become *incised* into the **flood plain**, and **river terraces** may be created.

relative humidity The relationship between the actual amount of water vapour in the air and the amount of vapour the air could hold at a particular temperature. This is usually expressed as a percentage. Relative humidity gives a measure of dampness in the **atmosphere**, and this can be determined by a **hygrometer**.

relief The differences in height between any parts of the Earth's surface. Hence a relief map will aim to show differences in the height of land by, for example, **contour** lines or by a colour key.

relief rainfall See **orographic rainfall**.

renewable resources Resources that can be used repeatedly, given appropriate management and conservation. Water and timber are examples of these. Solar energy and wind energy are also renewable resources. Compare **nonrenewable resources**.

representative fraction The fraction of real

size to which objects are reduced on a map; for example, on a 1:50 000 map, any object is shown at 1/50 000 of its real size.

reserves Resources which are available for future use. The world has reserves of **fossil fuels** which, if used with careful management, could last for many years. However, many of our fossil fuel and mineral reserves have been used up at a rapid rate by the developed countries, and more controlled use is required in the future to conserve reserves.

resource Any aspect of the human and physical **environments** which people find useful in satisfying their needs.

respiration The release of energy from food in the cells of all living organisms (plants as well as animals). The process normally requires oxygen and releases carbon dioxide. It is balanced by **photosynthesis**.

retail industry The people and organizations involved in the selling of goods to the public, usually in shops. Goods coming from manufacturers are distributed to the retail industry by *wholesalers*. A *retail park* is a large shopping centre built on the edge of a town to which consumers travel by car.

revolution The passage of the Earth around the sun; one revolution is completed in 365.25 days. Due to the tilt of the Earth's axis ($23\frac{1}{2}°$ from the vertical), revolution results in the sequence of seasons experienced on the Earth's surface.

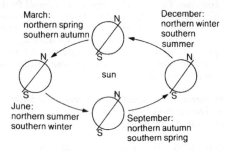

March:
northern spring
southern autumn

December:
northern winter
southern summer

sun

June:
northern summer
southern winter

September:
northern autumn
southern spring

Revolution The seasons of the year.

ria A submerged river valley, caused by a rise in sea level or a subsidence of the land relative to the sea. For example, a postglacial rise in sea level may lead to coastal submergence.

In this example, a rise in sea level of 25 metres would create a ria. Note the winding plan; contrast with **fjord**.

The seaboard of southwest Ireland is a classic ria coastline. See also **discordant coastline**.

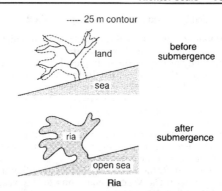

----- 25 m contour

land

sea

before submergence

ria

open sea

after submergence

Ria

ribbon lake A long, relatively narrow lake, usually occupying the floor of a U-shaped glaciated valley. A ribbon lake may be caused by the *overdeepening* of a section of the valley floor by glacial **abrasion**. Such a situation would occur, for example, where softer or weakened **rocks** outcrop on the valley floor. Alternatively, a ribbon lake may be dammed back by a terminal or recessional **moraine**. Many of the lakes of the English Lake District – for example Windermere, Coniston Water, Wastwater – are ribbon lakes.

Richter scale A scale of **earthquake** measurement that describes the magnitude of an earthquake according to the amount of energy released, as recorded by **seismographs**. See **Appendix 2**.

rift valley A section of the Earth's **crust** which has been downfaulted. The **faults** bordering the rift valley are approximately parallel. There are two main theories related to the origin of rift valleys. The first states that tensional forces within the Earth's **crust** have caused a block of land to sink between parallel faults.

The second theory states that compression within the Earth's crust has caused faulting in which two side blocks have risen up towards each other over a central block.

The river Rhine flows through a rift valley in Europe, with the Vosges mountains to the west and the Black Forest to the east.

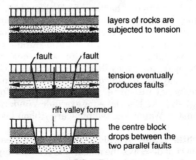

layers of rocks are subjected to tension

tension eventually produces faults

fault   fault

rift valley formed

the centre block drops between the two parallel faults

Rift valley

The most complex rift valley system in the world is that ranging from Syria in the Middle East to the river Zambesi in East Africa.

river basin   The area drained by a river and its tributaries, sometimes referred to as a **catchment** area. See diagram below.

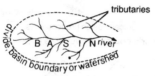

River basin

river cliff or **bluff**   The outer bank of a **meander**. The cliff is kept steep by undercutting since river **erosion** is concentrated on the outer bank. See **meander** and **river's course**.

river's course   The route taken by a river from its source to the sea. There are three major sections: the upper course, the middle course and the lower course. The characteristics of these stages are shown in the diagrams overleaf.

river terrace   A platform of land beside a river. This is produced when a river is **rejuvenated** in its middle or lower courses. The river cuts down

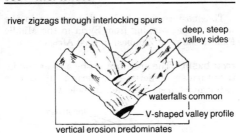

river zigzags through interlocking spurs

deep, steep
valley sides

waterfalls common

V-shaped valley profile

vertical erosion predominates

River's course A river in its upper course.

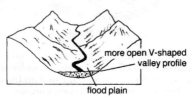

more open V-shaped
valley profile

flood plain

River's course A river in its middle course.

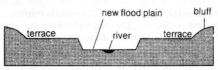

new flood plain

bluff

terrace

river

terrace

River terrace Paired river terraces above a flood plain.

river bluffs where spurs have been removed

wide floodplain

oxbow lake

levées

lateral erosion predominates

thick alluvial deposits

shallow, flat-bottomed valley profile

River's course A river in its lower course.

into its **flood plain**, which then stands above the new general level of the river as paired terraces. If a river on a flood plain is rejuvenated several times, a series of paired terraces can develop.

roche moutonnée An outcrop of resistant **rock** sculpted by the passage of a **glacier**.

passage of ice ⟶

upstream side smoothed and striated by abrasion

downstream side jagged due to plucking

chattermarks – hollows chiselled out by rocks embedded in ice

Roche moutonnée

rock The solid material of the Earth's **crust**. See **igneous rock, sedimentary rock, metamorphic rock**.

rotation The movement of the Earth about its own axis. One rotation is completed in 24 hours. Due to the tilt of the Earth's axis, the length of day and night varies at different points on the Earth's surface. In the northern midsummer, for example, the situation illustrated in the diagram

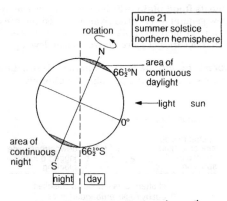

Rotation The tilt of the Earth at the northern summer and southern winter solstice.

prevails. At the equator there is a 12-hour day and a 12-hour night. North of $66\frac{1}{2}°$N there is continuous daylight; south of $66\frac{1}{2}°$S there is continuous night. Days become longer with increasing latitude north; shorter with increasing latitude south.

route   The course taken between starting point and destination. For example, the route of the M1 motorway from London to Leeds is via Leicester and Sheffield. The study of routeways can be important to geographers in establishing the **accessibility** of an area or examining the effects of a new road on the **environment**.

rural depopulation   The loss of population from the countryside as people move away from rural areas towards cities and **conurbations**. Most industrial nations have experienced a shift of population from the countryside to the city as the economy has evolved from an agricultural into a predominantly urban-industrial system. Rural communities tend to lose their most dynamic members as, generally, the migrating population consists chiefly of the younger, most active, most progressive individuals.

rural–urban migration   The movement of people from rural to urban areas. See **migration** and **rural depopulation**.

sandstone A common **sedimentary rock** deposited by either wind or water.

Sandstones vary in texture from fine to coarse grained, but are invariably composed of grains of **quartz**, cemented by such substances as calcium carbonate or silica.

satellite image An image giving information about an area of the Earth or another planet, obtained from a satellite. Instruments on an Earth-orbiting satellite, such as Landsat, continually scan the Earth and sense the brightness of reflected light. When the information is sent back to Earth, computers turn it into *false colour images* in which built-up areas appear in one colour (perhaps blue), vegetation in another (often red), bare ground in a third, and water in a fourth colour, making it easy to see their distribution and to monitor any changes, such as **deforestation**. Satellite photography can be used to measure the depth of shallow water, to monitor the health and development of forestry and agricultural crops, and as an aid in oil and mineral exploration.

savanna The grassland regions of Africa which lie between the **tropical rain forest** and the hot **deserts**. Thus, in the states of West Africa, for example, a range of vegetation exists from south to north: tropical rainforest to hot desert via a transition zone of savanna grassland.

Near the forest margin the savanna region comprises extensive woodland with occasional expanses of grassland; near the desert margin the vegetation is sparse, thorny scrub. Commercial **pastoral farming** is the dominant land use of the central savannas, while **nomadic pastoralism** prevails towards the desert margin. In South America the *Llanos* and *Campos* regions are representative of the savanna type.

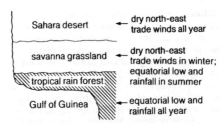

Savanna The position of the savanna in West Africa.

scale The size ratio represented by a map; for example, on a map of scale 1:25 000 the real landscape is portrayed at 1/25 000 of its actual size.

scarp slope The steeper of the two slopes which comprise an **escarpment** of inclined

strata. Compare **dip slope**.

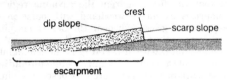

Scarp slope

scatter graph   A graph showing the scores of a set of individuals or items on two variables. The *line of best fit* runs through the points so that the sum of the squares of the vertical distances (or offsets) from the points on to the line is minimal (see figure (a)). The purpose of the graph is to indicate visually the general trend of the data, i.e. the degree of relationship, or correlation, between the two variables applying to that set of individuals or items. When a unique position for the line cannot be found, because of the purely random distribution of points, as shown in figure (b), then no relationship exists and the correlation value is zero.

   The correlation may be positive: the scores on both variables increase or decrease together (figure (a)); or negative: the scores on one of the variables increase while those of the other decrease (figure (c)).

Displays of points may take many other forms. For example, figure (d) shows clustering in which the correlation of each cluster is zero, but all the individuals of each have common characteristics not shared with those of the other clusters.

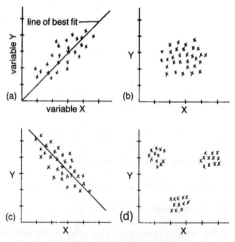

Scatter graph (a) Positive correlation. (b) Zero correlation. (c) Negative correlation. (d) Clustering.

science park A site accommodating several companies involved in scientific work or research. Science parks are linked to universities and tend to be located on 'greenfield' and/or landscaped sites. Compare **business park**.

scree or **talus** The accumulated **weathering** debris below a **crag** or other exposed rock face. Larger boulders will accumulate at the base of the scree, carried there by greater momentum.

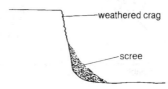

Scree The formation of scree.

sea breeze See **land breeze**.

sea-floor spreading See **plate tectonics**.

secondary sector The sector of the economy which comprises manufacturing and processing industries, in contrast with the **primary sector** which produces **raw materials**, the **tertiary sector** which provides **services**, and the **quaternary sector** which provides information.

An example of the secondary sector is the **iron and steel industry**. See **heavy industry**.

secondary source   A supply of information or data that has been researched or collected by an individual or group of people and made available for others to use; **census** data is an example of this. A *primary source* of data or information is one collected at first hand by the researcher who needs it; for example, a traffic count in an area, undertaken by a student for his or her own project.

sector model or **Hoyt model**   A model of urban structure developed by H. Hoyt in 1939 and based on an analysis of the land-use patterns of 142 American cities.

The model differs from the **Burgess model** in allowing sectorial development along major lines of communication. Thus, zone 2 (industry) and its related workers' housing zone (3) will develop along, for example, a river valley which favoured canal and railway expansion. Zone 5, high-quality housing, will extend along ridges of high ground or other pleasant environmental corridors. Whilst Hoyt's model is a step closer to reality, the **multiple nuclei model** is more so. No model will ever be an entirely successful simulation of real urban land-use patterns, as the local factors responsible for the structure of a given city will be unique to that location. Compare **shanty town**.

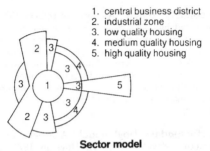

1. central business district
2. industrial zone
3. low quality housing
4. medium quality housing
5. high quality housing

Sector model

sediment The material resulting from the **weathering** and **erosion** of the landscape, which has been deposited by water, ice or wind. It may be reconsolidated to form **sedimentary rock**.

sedimentary rock A rock which has been formed by the consolidation of **sediment** derived from pre-existing rocks. **Sandstone** is a common example of a rock formed in this way; mudstone and shale are other examples. Such sedimentary rocks often show evidence of **bedding planes** which differentiate the layers of **deposition** sequences. **Chalk, limestone** and **evaporites** are other types of sedimentary rock, derived from organic and chemical precipitations.

seif dune A linear sand dune, the ridge of sand lying parallel to the prevailing wind direction. The eddying movement of the wind keeps the sides of the dune steep. Seif dunes can be up to 200 m high and can run for over 10 km.

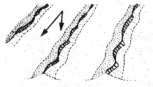

Seif dunes

seismograph An instrument that measures and records the **seismic waves** which travel through the Earth during an **earthquake**.

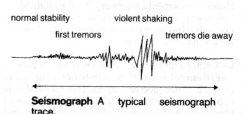

normal stability   violent shaking

first tremors   tremors die away

Seismograph A typical seismograph trace.

The energy in these waves is measured on the **Richter scale**.

seismology The study of **earthquakes**.

serac A pinnacle of ice formed by the tumbling and shearing of a **glacier** at an **ice fall**, i.e. the broken ice associated with a change in **gradient** of the valley floor.

service industry The people and organizations that provide a service to the public. Examples are the transport industry, banking, the health service, the education service and the **retail industry**. The largest concentrations of service industries are found in urban areas. Service industries are in the **tertiary sector**.

settlement Any location chosen by people as a permanent or semipermanent dwelling place. Hence, settlements may vary from an individual farmhouse in an agricultural landscape, to a **conurbation** of several millions of people in an urban/industrial region.

settlement hierarchy A series of size orders in a **settlement** system, the number of settlements in each order decreasing as the hierarchy is ascended. Generally, a settlement hierarchy will comprise many low-order places (or villages) and very few high-order places (or cities), with a number of intermediate orders between the two extremes. The logic behind the hierarchy concept is formalized in Walter Christaller's **central place theory**. Christaller suggested 'k' values to indicate the numerical relationship between one order and another; for example, in his k=3 hierarchy the number of settlements is three times fewer at each successive higher order.

Such a hierarchy is, of course, entirely theoretical, but empirical studies do suggest some evidence of hierarchical organization.

Order	Number
4th order:	1
3rd order:	3
2nd order:	9
1st order:	27

Settlement hierarchy

shading map or **choropleth map**   A map in which shading of varying intensity is used. For example, the pattern of **population densities** in a region can be shown by means of a shading system (see diagram overleaf).

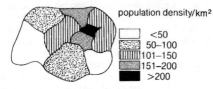

population density/km²

☐ <50
50–100
101–150
151–200
■ >200

Shading map

shaft mine A vertical mine. Compare **drift mine**.

shanty town An area of unplanned, random, urban development often around the edge of a city and generally occurring in **Third World** nations – cities such as São Paulo, Mexico City, Nairobi, Calcutta and Lagos have shanty towns. The shanty town is characterized by high-density/low-quality dwellings, often constructed from the simplest materials such as scrap wood, corrugated iron and plastic sheeting – and by the lack of standard services such as sewerage and water supply, power supplies and refuse collection. The shanty town is the makeshift home of the relatively recent rural-urban migrant: unemployment tends to be high since the supply of urban employment is considerably less than the demand. However, the conventional image of the shanty town is misleading: the fact that such **settlements** survive is a measure of success; many occupants gradually improve their property as they become established in the urban sector, and the **informal economy** thrives.

The shanty town is a major element of the structure of many Third World cities. Shanty towns in Brazil are known as *favelas*, in Calcutta as *bustees*. A typical land-use pattern of a Third World city is given in the diagram. Compare this with models for first-world cities – **Burgess model, sector model** and **multiple nuclei model**.

Shanty town A model of a typical Third World city.

shifting cultivation   See **bush fallowing**.

shoreface terrace   A bank of **sediment** accumulating at the change of slope which marks the limit of a marine **wave-cut platform**.

Material removed from the retreating cliff base is transported by the **undertow** off the wave-cut platform to be deposited in deeper water offshore. See diagram below.

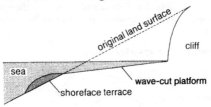

Shoreface terrace

silage   Any **fodder crop** harvested whilst still green. The crop is kept succulent by partial fermentation in a *silo*. It is used as animal feed during the winter.

sill   1. An igneous intrusion of roughly horizontal disposition. Some sills form large ridge-like escarpments when exposed; an example is the Great Whin Sill in northern England.

2. (also called **threshold**) The lip of a **corrie**. See diagram.

corrie

sill or threshold; maximum ice thickness was at centre of corrie which was therefore eroded to greater depth than the sill.

mountainside

Sill The sill of a corrie.

silt Fine **sediment**, the component particles of which have a mean diameter of between 0.002 mm and 0.02 mm.

sinkhole See **pothole**.

slash and burn See **tropical rainforest**.

slate Metamorphosed shale or **clay**. Slate is a dense, fine-grained **rock** distinguished by the characteristic of *perfect cleavage*, i.e. it can be split along a perfectly smooth plane.

slip The amount of vertical displacement of **strata** at a **fault**.

slip-off slope The relatively gentle slope opposite a river cliff or **bluff** at a river **meander**. In the valley of a meandering river, the route of the fastest flowing water will undercut the outer bank of the meander. Eroded material from this bank may be deposited on the inner bank of the meander where the water is flowing at a slower speed. This deposited material forms a convex bank called a 'slip-off slope'.

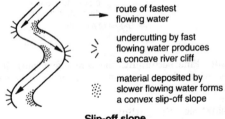

route of fastest flowing water

undercutting by fast flowing water produces a concave river cliff

material deposited by slower flowing water forms a convex slip-off slope

Slip-off slope

smog A mixture of smoke and fog associated with urban and industrial areas, that creates an unhealthy **atmosphere**. In 1952 a four-day smog in London left 4000 people dead or dying. Since then, many cities have had to introduce *clean air zones*. Los Angeles, in the USA, is renowned for its photochemical 'smog' caused by the effect of the sun's heat on car exhaust emissions.

smokestack industry A term used to describe any traditional heavy industry – iron and steel, shipbuilding – as opposed to newer, cleaner industries such as electronics.

snout The end of a **glacier**; strictly the point at which wasting by melting exceeds the rate of supply of ice from up-valley. A glacier snout is characterized by a dissected, discoloured appearance caused by the action of meltwater and the washing out of **moraine**.

snow line The altitude above which permanent snow exists, and below which any snow that falls will not persist during the summer months. The altitude of the snow line varies with **latitude**: at the equator it is approximately 5000 metres, in the Alps it is approximately 3000 metres, and at the Poles the snow line is at sea level.

socioeconomic group A group defined by particular social and economic characteristics, such as educational qualifications, type of job, and earnings. In **census** data, the socioeconomic groups are usually as follows: professional; employers and managers; non-manual; skilled workers; semi-skilled workers; and unskilled workers.

soil The loose material which forms the uppermost layer of the Earth's surface, composed of the **inorganic fraction**, i.e. material derived from the **weathering** of bedrock, and the **organic fraction** – that is material derived from the decay of vegetable matter. The ultimate form of the organic fraction is humus. Soil also contains minerals and trace elements, and air and water held within the soil structure. See **horizon**. There are three broad categories of soil: *zonal*, i.e. the result of a specific combination of climatic and vegetative conditions; *intrazonal*, i.e. the result of local peculiarities of **environment** such as waterlogging or unusual parent material; and *azonal*, i.e. little-developed soils which occur on such surfaces as **scree, alluvium** or sand dunes.

soil erosion The accelerated breakdown and removal of soil due to poor management. Soil erosion is particularly a problem in harsh **environments**. For example, in areas of steep slope, heavy seasonal rainfall, or strong dry-season winds, the soil may be rapidly eroded once protective vegetation has been removed. See **rainfall interception**. Severe soil erosion can cause problems of flooding and the silting-up of water supplies, especially in reservoirs. Erosion causes soil, one of our most valuable **resources**, to be lost, much of it ending up in the sea.

Soil erosion by rainwater can be classified as follows: *splash erosion*, whereby the soil is

pulverized by the impact of heavy raindrops and
hailstones, as in a convectional storm; *sheet
erosion*, whereby, in a heavy storm, a surface film
of water develops which will flow downslope
carrying surface soil as it moves; and *rill/gully
erosion*, whereby any surface depression
concentrates run-off which quickly develops into
channel flow, cutting a steep-sided valley as it
runs off. Soil erosion by wind occurs on extensive

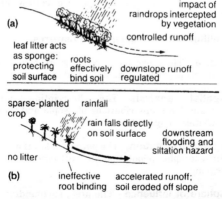

Soil erosion (a) Stable environment. (b)
Unstable environment.

flatlands which are subject to a windy, dry season for part of the year. The upper soil surface becomes loose and susceptible to wind erosion due to lack of moisture.

soil profile   The sequence of layers or **horizons** usually seen in an exposed soil section.

solar power   Heat radiation from the sun converted into electricity or used directly to provide heating. Solar power is an example of a renewable source of energy (see **renewable resources**).

solifluction   See **periglacial features**.

space   The geographer's term for area; the context within which distributions and patterns occur.

spatial analysis   The description and explanation of distributions of people and their activities in **space**.

spatial distribution   The pattern of locations of, for example, population or **settlement** in a region.

sphere of influence   The area, surrounding a **settlement**, from which consumers will travel in order to obtain goods and **services**. The size of the sphere of influence will vary according to the size of the settlement and the number and type of functions available there. For example, a neighbourhood shopping centre will have a *small* sphere of influence, i.e. people will not travel great distances to the centre as the range of goods and services will be limited. The central business district (**CBD**) of a town or city, however, will have a *large* sphere of influence. This means that people will travel from greater distances to the CBD, as the range of goods and services is much wider.

spit A low, narrow bank of sand and shingle built out into an **estuary** by the process of **longshore drift**. Spurn Head, Humberside, is a classic example. The **sediment** carried down the coast by longshore drift is deposited at the break in the coastline caused by the Humber estuary.

Spit Spurn Head, a coastal spit.

Relatively shallow water, and the slowing of longshore drift by the counter-current of the Humber, have led to the **deposition** of the marine **load**.

The end of Spurn Head is 'hooked' by the action of waves swinging into the Humber estuary from the open North Sea.

spring The emergence of an underground stream at the surface, often occurring where **impermeable rock** underlies **permeable rock** or **pervious rock** or **strata**.

permeable limestone
underground stream
spring
impermeable slate

Spring Rainwater enters through the fissures of the limestone and the stream springs out where the limestone meets slate.

spring line A series of **springs** emerging along the foot of an **escarpment** of **permeable rock** or **pervious rock**. *Spring line settlements* can occur at these points, tapping the natural water supply. The settlements along the chalk escarpments of southern England are an example.

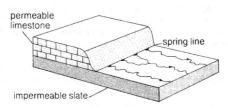

Spring line

spring tides See **tides**.

squatter settlement An area of peripheral urban settlement in which the residents occupy land to which they have no legal title. See **shanty town**.

stack A coastal feature resulting from the

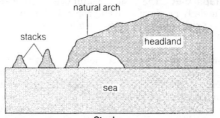

Stack

collapse of a **natural arch**. The stack represents the more resistant **strata**, while softer material has been worn away by **weathering** and marine **erosion**.

stalactite   A column of calcium carbonate hanging from the roof of a **limestone** cavern. As water passes through the limestone it dissolves a certain proportion, which is then precipitated by **evaporation** of water droplets dripping from the cavern roof. The drops splashing on the floor of a cavern further evaporate to precipitate more calcium carbonate as a **stalagmite**.

stalagmite   A column of calcium carbonate growing from a cavern floor. Compare **stalactite**. Stalactites and stalagmites may meet, forming a column or pillar.

staple diet   The basic foodstuff which comprises the daily meals of a given people. In South East Asia rice is the staple food, while, in many parts of Africa the staple food is maize.

Stevenson's screen   A shelter used in weather stations, in which thermometers are hung. The screen consists of a wooden box painted white and raised just over a metre from the ground. The four sides are louvred to allow free entry of air, and the roof is of double thickness to prevent the sun's rays reaching inside.

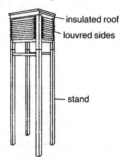

insulated roof

louvred sides

stand

Stevenson's screen

SSSI *abbreviation for* Site of Special Scientific Interest, an area containing rare species of plants or animals, or features of special geological interest.

SSSIs are designated in England by English Nature, but are mostly privately owned. English Nature has to be informed of developments being planned for SSSIs and can advise on their management. There are over 4000 SSSIs in England, accounting for 7% of the total land area. Threats to SSSIs come from neglect, overgrazing and development.

storm hydrograph A graph showing the **discharge** of a stream and its response to a period of rainfall. There is a time lag between peak rainfall and peak discharge. This is because it takes time for the rainwater to pass through the various 'storage units', i.e. vegetation, soil and rock. A series of storm hydrographs can be plotted over several days to give a picture of the river's *regime* or pattern.

strata   Layers of **rock** superimposed one upon the other; thus a sequence of strata is referred to as the *stratigraphy* of a region. The study of stratigraphy enables scientists to reconstruct the geological history of a region; this may be a complex process if the stratigraphy has been disturbed by earth movements such as folding or faulting.

stratosphere   The layer of the **atmosphere**

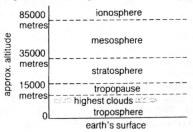

Stratosphere Structure of the atmosphere.

which lies immediately above the troposphere and below the mesosphere and ionosphere. Within the stratosphere, temperatures increase with altitude. The boundary between the stratosphere and the troposphere is known as the *tropopause*.

stratus Layer-cloud of uniform grey appearance, often associated with the warm sector of a **depression**. Stratus is a type of low **cloud** which may hang as mist over mountain tops; broken stratus is referred to as *fractostratus*.

striations The grooves and scratches left on bare **rock** surfaces by the passage of a **glacier**. Debris embedded in the glacier scores the surface over which it passes. Striations may thus be a guide to the direction of ice movement.

strip cropping A method of **soil** conservation whereby different crops are planted in a series of strips, often following **contours** around a hillside. The purpose of such a sequence of cultivation is to arrest the downslope movement of soil, especially if one of the crops (e.g. maize) tends to expose the soil surface. Alternate strips may be planted with grass which effectively binds the soil and acts as a brake on run-off. See **soil erosion**.

subduction zone   See **plate tectonics**.

subsoil   See **soil profile**.

subsidiary cone   A volcanic cone which develops within the **caldera** of a previous, larger cone which was blown away during an eruption.

subsistence agriculture   A system of **agriculture** in which farmers produce exclusively for their own consumption, in contrast to **commercial agriculture** where farmers produce purely for sale at the market.

suburbs   The outer, and largest, parts of a town or city. They consist mainly of low-density housing (either privately owned houses or council houses owned by the local authority). As the suburbs have spread outwards into the surrounding countryside, they have sometimes engulfed formerly rural villages. Suburbs also contain shopping centres, open spaces, schools, leisure facilities, etc. Offices, warehouses and small firms may also locate in the suburbs, taking advantage of the lower land values compared with the **CBD**. Beyond the suburbs, many towns and cities have a **green belt** to stop **urban sprawl**.

sunshine recorder   An instrument used to

measure the hours of sunshine in an area. It consists of a glass sphere mounted on a frame. The sun's rays are concentrated by the sphere onto a strip of sensitized paper behind. This paper is graduated into hours. When the sun shines, a line is burnt along the paper as the Earth rotates. The total length of the burnt line will thus indicate how many hours in the day the sun has shone. On a meteorological map, parallel lines drawn through places having the same amount of sunshine are called *isohels*.

surface run-off That proportion of rainfall received at the Earth's surface which runs off either as channel flow or overland flow. It is distinguished from the rest of the rainfall, which either percolates into the soil or evaporates back into the **atmosphere**.

sustainable development The ability of a country to maintain a level of economic development, thus enabling the majority of the population to have a reasonable standard of living.

swallow hole See **pothole**.

swash The rush of water up the beach as a wave breaks. See also **backwash** and **long-shore drift**.

syncline A trough in folded **strata**; the opposite of **anticline**. See **fold**.

taiga The extensive **coniferous forests** of Siberia and Canada, lying immediately south of the arctic **tundra**. Within the taiga there are many lakes, marshes and swamps, the latter often resulting from springtime thaw occurring over permafrost and while river courses to the north are still frozen. See also **periglacial features**.

tarn The postglacial lake which often occupies a **corrie**. See diagram below.

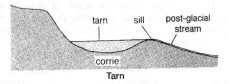

Tarn

temperate climate A climate typical of mid-latitudes. The British Isles, for example, have a temperate climate. Such a climate is intermediate between the extremes of hot (tropical) and cold (polar) climates. Compare **extreme climate**. See also **maritime climate**.

terminal moraine See **moraine**.

terrace 1. See **river terrace**.

2. A type of housing in which all the dwelling units are joined together, as opposed to detached or semi-detached housing.

terracing A means of **soil** conservation and land utilization whereby steep hillsides are engineered into a series of flat ledges which can be used for **agriculture**, held in places by stone banks to prevent **soil erosion**. Terracing reaches its most sophisticated and intricate form in the fertile volcanic hills of Java, Indonesia, where a subsistence rice economy prevails.

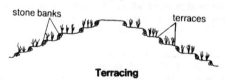

Terracing

tertiary sector That sector of the economy which provides **services** such as transport, finance and retailing, as opposed to the **primary sector** which provides **raw materials**, the **secondary sector** which processes and manufactures products, and the **quaternary sector** which provides information and expertise. In highly developed economies the tertiary sector is the dominant employer, though in recent years numbers employed in the quaternary sector have gone up due to increased use of computers and electronic information.

textile industry   The manufacture of cloth and related materials. Traditionally the textile industry used wool, cotton, flax and other natural products as its **raw materials**, but this sector has declined with the rise in use of **artificial fibres** derived from oil and other hydrocarbons. Nylon, rayon and polyester are examples of such fibres

thermal power station   An electricity-generating plant which burns **coal**, oil or natural gas to produce steam to drive turbines.

Third World   A collective term for the poor nations of Africa, Asia and Latin America, as opposed to the 'first world' of capitalist, developed nations and the 'second world' of formerly communist, developed nations. The terminology is far from satisfactory as there are great social and political variations within the 'Third World'. Indeed, there are some countries where such extreme poverty prevails that these could be regarded as a fourth group. Alternative terminology includes **'developing countries'**, 'economically developing countries' and 'less economically developed countries' (LEDC).

Newly industrialized countries are those showing greatest economic development.

threshold   See **sill** (sense 2).

threshold population   The minimum population required in a **sphere of influence** to sustain a particular good or service offered in a **central place**. The size of the threshold population increases with progressively more specialized goods and services for which individual demand is less frequent.

tidal limit   The point upstream of which there is no tidal rise and fall in river level.

tidal range   The mean difference in water level between high and low tides at a given location. See **tides**.

tides   The alternate rise and fall of the surface

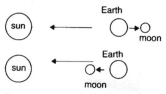

Tides Spring tide.

of the sea, approximately twice a day, caused by the gravitational pull of the moon and, to a lesser extent, of the sun.

Tides Neap tide.

When the moon, earth and sun are in a straight line, at full and new moon, the gravitational force is at its greatest because the pull of the sun is combined with that of the moon. This produces a *spring tide*, characterized by a large tidal range (very high tides and equally low tides). When the moon, earth and sun are not in a straight line, the effect on the Earth is less. The pull of the sun and of the moon are at

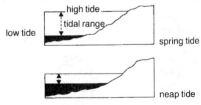

Tides Tidal ranges.

right angles when there is a half moon. At this time, the difference between high and low tides is not very great. These tides are called *neap tides*.

till See **boulder clay**.

tombolo A **spit** which extends to join an island to the mainland, as in the case of Chesil Beach, Portland Island, southern England.

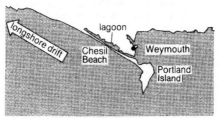

Tombolo Chesil Beach.

topography The composition of the visible landscape, comprising both physical features (e.g. relief, **drainage** and vegetation) and those made by people (e.g. roads, railways and **settlements**).

topsoil The uppermost layer of **soil**, more rich in organic matter than the underlying **subsoil**.

See **horizon**, **soil profile**.

tourist industry The people and organizations that provide facilities and services for tourists. It is a major industry in many countries of the world, employing millions of people, either directly or indirectly.

In some developing countries, a tourist industry has brought in capital, but tourism on a large scale can have detrimental effects on the traditional culture and lifestyle of the indigenous population, and extensive hotel-building damages the countryside. Some countries, such as Tanzania, have developed a form of eco-tourism designed to encourage specialized visitors, rather than mass-market tourism, as a way of protecting the environment. Large numbers of tourists to any area can cause problems, ranging from footpath erosion in a **national park** such as the Lake District, to congested roads and air **pollution** by motor vehicles. Tourism and **conservation** can come into conflict over the destruction either of the natural environment or of a city such as Venice.

trade winds Winds which blow from the subtropical belts of high pressure towards the equatorial belt of low pressure. In the northern hemisphere the winds blow from the northeast and in the southern hemisphere from the southeast. In most areas these winds blow with great regularity, and weather in the trade wind regions is usually fine and quiet. However, these regions can sometimes experience **hurricanes**.

transhumance   The practice whereby herds of farm animals are moved between regions of different climates. Pastoral farmers (see **pastoral farming**) take their herds from valley pastures in the winter to mountain pastures in the summer. Frequently farmers will live in mountain huts during the summer in mountainous regions such as the Himalayas and the Alps.

transnational corporation (TNCs)   A company that has branches in many countries of the world, and often controls the production of the primary product and the scale of the finished article. For example, TNCs own many tea plantations in the Third World, where the tea is picked and processed by local labour (see **plantation agriculture**). The TNCs also control the selling of the tea, most of which is sent to developed countries. It is argued by some economists that labour in the Third World is badly paid in order to provide cheap products worldwide. The TNCs, however, claim to provide higher wages and better working conditions than indigenous industries.

transpiration   The process whereby plants give off water vapour via the stomata of their leaves. Water taken up by roots is thus returned to the **atmosphere**.

tropical rainforest   The dense forest cover of the equatorial regions, reaching its greatest extent in the Amazon Basin of South America, the Congo Basin of Africa, and in parts of South East Asia and Indonesia. The lush forest is a response to optimum climatic conditions (high temperatures and abundant moisture), and not to soil fertility. Tropical soils are, in fact, generally poor. Newly germinated seedlings grow rapidly upwards in search of light in the dense forest cover, before branching out having reached the forest canopy. Most trees are very shallow rooted, and are often buttressed to provide support. The richness and diversity of the forest itself gives rise to a similarly varied fauna. There has been much concern in recent years about the rate at which the world's rainforests are being cut down. There are several reasons for this **deforestation**, such as *slash and burn* farming by the local population, and, more especially, large-scale forest clearance to provide land for cattle farming or for development projects such as **hydroelectric power** schemes and mineral excavation. The burning of large tracts of rainforest is thought to be contributing to **global warming**. Many governments and **conservation** bodies are now examining ways of protecting the remaining rainforests, which are unique **ecosystems** containing millions of plant and animal species.

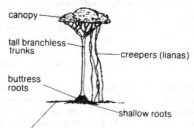

canopy

tall branchless trunks

creepers (lianas)

buttress roots

shallow roots

intense bacterial activity breaks down fallen leaves etc. to return nutrients to soil surface for immediate uptake by roots.
Soils themselves are infertile: the nutrient cycle is concentrated in the vegetation and top few inches of soil.

Tropical rainforest Typical characteristics.

trough An area of low pressure, not sufficiently well-defined to be regarded as a **depression**.

trough end wall The steep rear wall of a

U-shaped valley, formed where coalescing **corrie glaciers** cause an increase in **erosion** and a consequent deepening of the glacial valley.

truncated spur   A spur of land that previously projected into a valley and has been completely or partially cut off by a moving **glacier**. The removal of spurs in a river valley therefore has the effect of widening and deepening the valley. See **U-shaped valley**. The position of the truncated spur is usually marked by a **crag** and **scree**.

tsunami   A very large, and often destructive, sea wave produced by a submarine **earthquake**. Tsunamis tend to occur along the coasts of Japan and parts of the Pacific Ocean, and can be the cause of large numbers of deaths. Tsunamis are sometimes incorrectly referred to as 'tidal waves'.

tuff   Volcanic ash or dust which has been consolidated into **rock**.

tundra   The barren, often bare-rock plains of the far north of North America and Eurasia where subarctic conditions prevail and where, as a result, vegetation is restricted to low-growing, hardy shrubs and mosses and lichens. Permafrost conditions result in poor **drainage** which results in marsh and swamp during the short summer. See **periglacial features**.

undernutrition   A lack of a sufficient quantity of food, as distinct from **malnutrition** which is a consequence of an unbalanced diet. Many of the world's poorest people suffer both undernutrition and malnutrition. As the world's population rises at a faster rate than that of the food supply, problems of undernutrition are becoming worse in many parts of Latin America, Africa and Asia.

undertow   The counter-current to water breaking onshore as waves. The undertow is responsible for the removal of eroded material from the **wave-cut platform**, to be deposited as the **shoreface terrace**. The undertow operates at a much larger scale than **backwash**.

urban decay   The process of deterioration in the **infrastructure** of parts of the city – especially in the old industrial cities of, for example, Northern England and the Midlands.
   Parts of the inner city are especially decayed: old Victorian **terrace** housing, mills and other traditional industrial installations such as disused canals and warehouses. Much of the decay results from neglect, as the focus of the urban system moves away from these areas – for example, towards new peripheral locations for industry, newer housing in the outer suburbs, and new **communications** bypassing the city. Many cities have undertaken 'face-lift' schemes to improve the

decayed **environments**, either by demolition and landscaping, or by renovating old property for new uses. See **comprehensive redevelopment**.

urbanization The process by which a national population becomes predominantly urban through a **migration** of people from the countryside to cities, and a shift from agricultural to industrial employment. Urbanization is thus an important element of the **development process**. See also **migration**.

urban sprawl The growth in extent of an urban

late
(19.

low mobility, majority of urban population walked to work

mid
(20)

spreading city in response to developing public transport

2000

high mobility due to higher incomes and car ownership

Urban sprawl The sequence of growth of urban areas in the UK.

area in response to improvements in transport and rising incomes, both of which allow a greater physical separation of home and work. The sprawling outer suburbs of cities in **developed countries** today are a result of almost universal car ownership; in earlier years similar, though smaller, expansions of the urban area resulted from the development of the urban railway, the tram and the bus. See diagram opposite. Urban sprawl has caused towns to merge. See **conurbation**, also **suburb** and **green belt**.

U-shaped valley A glaciated valley,

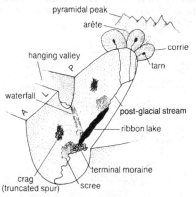

U-shaped valley The major features.

characteristically straight in plan and U-shaped in **cross section**.

The winding V-shaped valley of a river's course is modified into a U-shaped valley by the greater erosive power of a **glacier**.

vicious cycle of poverty The poverty trap in which much of the population of the **Third World** finds itself. Poor farmers cannot afford to invest in their land through improved seed or fertilizer: **yields** thus remain low, there is no surplus for sale at market, and so the poverty continues. In the absence of credit or grant aid, there is no way in which the farmer can break out of the cycle.

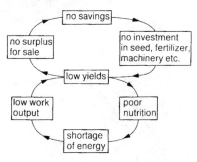

Vicious cycle of poverty

Given that poor yields also lead to a food shortage, the farmer will suffer poor nutrition and may even have to borrow money to secure enough food, let alone invest in the land. It is not only necessary to provide the financial means for the farmer to improve his or her livelihood, it is also important to provide the correct institutional context for progress – for example, **land reform** may be necessary to ensure security of tenure.

viscous lava **Lava** that resists the tendency to flow. It is sticky, flows slowly and congeals rapidly. *Nonviscous* lava is very fluid, flows quickly and congeals slowly.

volcanic rock A category of **igneous rock** which comprises those rocks formed from **magma** which has reached the Earth's surface. (This is to be contrasted with **plutonic rock** which forms below the surface.) **Basalt** is an example of a volcanic rock, as are all solidified lavas.

volcano A fissure in the Earth's **crust** through which **magma** reaches the Earth's surface. There are four main types of volcano:
(a) *Acid lava cone* – A very steep-sided cone composed entirely of acidic, **viscous lava** which flows slowly and congeals very quickly. The cones

of the Puy District in central France are classic examples.

(b) *Composite volcano* – A single cone comprising alternate layers of ash (or other pyroclasts) and lava. Such volcanoes as Vesuvius and Etna are of this type.

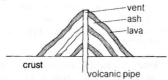

Volcano A composite volcano.

(c) *Fissure volcano* – A volcano that erupts along a linear fracture in the crust, rather than from a single cone. Fissure volcanoes are generally of a quiet, unexplosive nature, relating to the *non-viscous lavas* which are usually produced. Many eruptions in Iceland are of this type.

(d) *Shield volcano* – A volcano composed of very basic, nonviscous lava which flows quickly and congeals slowly, producing a very gently sloping

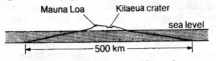

Volcano A shield volcano: Mauna Loa.

cone. Mauna Loa, Hawaii, is a good example of this type.

Von Thünen theory J. H. Von Thünen's early-19th century model of the distribution of agricultural land use around a town is still the basis for many investigations of such land-use patterns. Von Thünen envisaged a single market town, surrounded by a region supplying agricultural produce. Physical and economic characteristics of this agricultural area were regarded as uniform.

Such conditions of course do not exist in reality, but the value of a model like Von Thünen's is that it sweeps away the clutter of reality and allows us to observe basic processes at work. Von Thünen postulated the following land-use system, given his simplifying assumptions, illustrated in the diagram.

1 market gardening
2 dairying
3 arable
4 rough grazing
5 wilderness

Von Thünen theory Agricultural land use around a market town.

The zones 1-5 are derived from **bid-rent curves** thus:

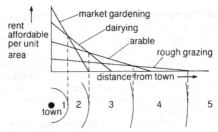

Von Thünen theory Bid-rent curves.

The implication of the theory is that the distribution of land uses around the town will depend upon a variety of factors: perishability, transport costs and land area requirements.

Highly perishable produce such as salad crops must be grown close to town to ensure freshness at market and to minimize the transport costs necessitated by frequent marketing. The location of market gardens close to the town is also explained by the fact that land area requirements are small, and therefore the bid rent per unit area can be high. Market gardeners can achieve high productivity from small areas by the intensive nature of their farming, and through investment in greenhouses, fertilizers and other inputs. Land uses with progressively larger land area requirements must locate further away from the town; a farmer needing many units of land will only be able to bid a lower rent per unit area. Land uses at progressively greater distance from town are also characterized by a decreasing frequency of marketing of products, and by decreasing in the land. In reality such patterns are constrained by variations in the physical **environment** and in human behaviour; government policy also affects farming practices and the distribution of land uses.

V-shaped valley   A narrow, steep-sided valley made by the rapid erosion of rock by streams and rivers. It is V-shaped in cross-section and is not a glacial feature. Compare **U-shaped valley**.

vulcanicity   A collective term for those processes which involve the intrusion of **magma** into the **crust**, or the extrusion of such molten material onto the Earth's surface.

wadi   A dry watercourse in an arid region; occasional rainstorms in the desert may cause a temporary stream to appear in a wadi.

warm front   See **depression**.

warm sector See **depression**.

waterfall An irregularity in the long profile of a **river's course**, usually located in the upper course. A waterfall occurs either at the edge of a stratum of resistant rock (**cap rock**), where it forms a lip, the softer rock immediately downstream being eroded more quickly than the resistant rock (see diagram); or where strata of resistant rock **dip** upstream (compare **rapid**): again the softer rock downstream wears away sooner, and a step is formed.

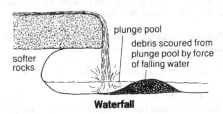

plunge pool

debris scoured from plunge pool by force of falling water

softer rocks

Waterfall

water gap In chalk **escarpments**, a valley which has been eroded below the depth of the **water table**, and which therefore contains a permanent stream. Compare **dry valley, wind gap**.

watershed The boundary, often a ridge of high

ground, between two **river basins**.

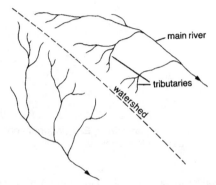

Watershed

water table   The level below which the ground is permanently saturated. The water table is thus the upper surface of the **groundwater**. In areas where **permeable rock** predominates, the water table may be at some considerable depth. In periods of high rainfall the level of the water table will rise; it will fall in protracted dry spells.

wave-cut platform or **abrasion platform**   A gently sloping bench eroded by the sea along a coastline. As the **cliff** recedes, the wave-cut

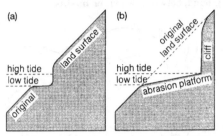

Wave-cut platform (a) Early in formation.
(b) Later in formation.

platform is created. The platform slopes seawards, since **erosion** has been active for a longer period of time at the seaward end. When sea levels are lowered, a wave-cut platform may appear as a *raised beach* several metres above the current sea level. Raised beaches can be seen on the western coast of the Kintyre peninsula in Scotland.

wave refraction   The bending of waves around a **headland**. The shorewards movement of water in contact with the headlands is slowed, while open water in the middle of the bay moves on unimpeded. See diagram opposite.

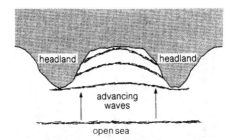

Wave refraction A vertical view.

weather The day-to-day conditions of e.g. rainfall, temperature and pressure, as experienced at a particular location. Contrast this with **climate**, which is a set of long-term, average atmospheric conditions. Thus the climate of, for example, a location in the Sahara **desert** can be characterized as hot and dry all the year round, but there will be occasions when violent weather in the form of convectional rainstorms occurs.

weather chart A map or chart of an area giving details of **weather** experienced at a particular time of day, such as 0600 hrs. The information is gathered at weather stations throughout the country, and when all observations have been

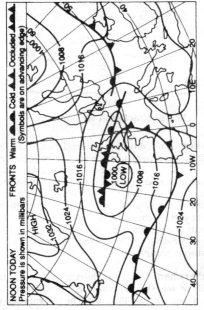

Weather chart A weather chart taken from a daily newspaper.

collected **isobars** can be drawn and the **depressions, anticyclones, fronts**, etc. can be identified. Weather charts are sometimes called *synoptic charts*, as they give a synopsis of the weather at a particular time.

weathering The breakdown of rocks *in situ*; contrasted with **erosion** in that no large-scale transport of the denuded material is involved. Weathering processes include **exfoliation, nivation** and chemical activity such as the dissolving of **limestone** by rainwater. Biological activity, for example by tree roots and earthworms, also contributes towards the breakdown of bedrock. Thus three types of weathering are identified: mechanical (or physical), chemical and biological.

weather station A place where all elements of the weather are measured and recorded. Each station will have a **Stevenson's screen** and a variety of instruments such as a **maximum and minimum thermometer**, a **hygrometer**, a **rain gauge**, a **wind vane**, an **anemometer** and a **sunshine recorder**.

Weber's theory See **industrial location**.

wet and dry bulb thermometer See **hygrometer**.

white-collar worker A worker who is not a

manual worker and who does not work in 'dirty conditions'. The term 'white-collar' derives from the white shirts worn by e.g. clerical workers and professional people whose jobs do not make their clothes dirty. Compare **blue-collar worker**.

white ice Ice from which air has not been totally expelled (see **corrie**). Contrast this with **blue ice** which is found at greater depth in a corrie icefield and from which air has been expelled by compression.

wind gap A **dry valley**, for example in a chalk **escarpment**, now standing above the **water table**, but formed at a time when the water table was higher or when the ground was frozen.

wind vane An instrument used to indicate wind direction. It consists of a rotating arm which always points in the direction from which the wind blows. The wind is named after this direction.

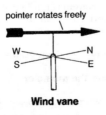

Wind vane

yield   The productivity of land as measured by the weight or volume of produce per unit area. Agricultural yields are usually expressed per hectare.

yardang   Long, roughly parallel ridges of **rock** in arid and semi-arid regions. The ridges are undercut by wind **erosion** and the corridors between them are swept clear of sand by the wind. The ridges are oriented in the direction of the prevailing wind. An area containing yardangs is often referred to as a 'ridge and furrow' landscape. The term 'yardang' comes from Central Asia.

Zeugen   *Pedestal rocks* in arid regions; wind **erosion** is concentrated near the ground where **corrasion** by wind-borne sand is most active. This leads to undercutting and the pedestal profile emerges.

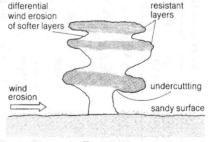

Zeugen

APPENDIX 1
Beaufort wind scale modified for use on land

SCALE FORCE	WIND NAME	SPEED in km/hr	WIND EFFECTS ON LAND
0	Calm Air	< 1	Calm; smoke rises vertically
1	Light Air	1–5	Direction of wind shown by smoke, but not by wind vanes
2	Light Breeze	6–11	Wind felt on face; leaves rustle; ordinary vane moved by wind
3	Gentle Breeze	12–19	Leaves and small twigs constantly moving; wind extends small flag
4	Moderate Breeze	20–29	Raises dust and loose paper; small branches are moved

5	Fresh Breeze	30–39	Small trees in leaf begin to sway; little wave crests form on inland waters
6	Strong Breeze	40–50	Large branches move; whistling heard in telephone wires, umbrellas used with difficulty
7	Near Gale	51–61	Whole trees move; inconvenience felt when walking against the wind
8	Gale	62–74	Breaks twigs off trees; progress generally hindered or impeded
9	Strong Gale	75–87	Slight structural damage occurs, such as roof slates or chimney pots removed
10	Storm	88–101	Considerable structural damage; trees uprooted; seldom experienced inland

| 11 | Violent Storm | 102–117 | Very rarely experienced; accompanied by widespread damage |
| 12 | Hurricane | over 118 | Countryside devastated |

APPENDIX 2
The Richter Scale

MAGNITUDE	EFFECTS AT EPICENTRE
1 to 3.4	None, only recorded on seismographs.
3.5 to 4.2	Feels like the vibrations due to a passing lorry. Noticed by people at rest.
4.3 to 4.8	Noticed by everyone. Loose objects are rocked, sleepers woken up and church bells ring.
4.9 to 5.4	Trees sway, loose objects fall. Some damage.
5.5 to 6.1	Walls crack, plaster falls. General alarm.
6.2 to 6.9	Some buildings collapse. Chimneys fall. Pipes break. Ground fissured or cracked.
7.0 to 7.3	Ground cracks badly. Buildings destroyed. Railway lines bent. Landslides.

| 7.4 to 8.1 | Only few buildings withstand shock, bridges destroyed. All pipes and cables broken, landslides, floods. |
| over 8.1 | Countryside devastated. Total destruction. |

APPENDIX 3
Study areas

The following countries are included in this Appendix:

Australia
Bangladesh
Brazil
China
Egypt
France
Germany
Ghana
India
Italy
Japan
Kenya
Mexico
Nigeria
Pakistan
Peru
Spain
United Kingdom
USA
Venezuela

AUSTRALIA

Physical background and climate
A country in the south Pacific comprising the whole of the smallest continent and the island of Tasmania to the southeast. Australia is the world's flattest and lowest land mass. Almost 75% of the continent is a vast ancient plateau. To the east is a highland belt stretching along the coast. There are several types of climate in Australia. The north, within the tropics, experiences high temperatures and heavy rainfall. 60% of the country lies within a temperate zone with warm summers and mild winters. Central Australia is semi-arid.

Area and population
Area – 7 687 000 sq. km
Population – 18 057 000
Population density – 2 per sq. km
Birth rate – 16 per 1000
Death rate – 8 per 1000
Annual rate of population growth – 1.4%
Urban population – 85%
Capital city – Canberra

Economy
Agriculture continues to make a valuable contribution to the economy, the main crops being cereals, sugar cane, wine and fruit.

Livestock, particularly sheep and cattle, are important. Mining is now of vital importance, especially the extraction of oil, coal, iron ore, bauxite, uranium, gold, diamonds, nickel, copper, lead and zinc.

Since the Second World War, Australia's manufacturing sector has expanded rapidly. In addition to steel, engineering, paper, chemicals, foodstuffs and textiles are important.

GDP – 392 600 million US dollars

GDP per capita – 21 740 US dollars

BANGLADESH

Physical background and climate
A country of the Indian subcontinent, having
borders with India and Myanmar (Burma).
Bangladesh is situated in the world's largest
delta, that of the rivers Ganges and
Brahmaputra. This vast alluvial plain in places
is very vulnerable to flooding. Swampy islands of
vegetation fringe the coastline. The only upland
areas of Bangladesh are the Chittagong Hills in
the southeast, and the Sylhet tea-growing area
in the northeast.

The climate of Bangladesh is monsoonal.
During the summer monsoon, May to October,
Bangladesh receives 80% of its annual rainfall.
Winter, from November to March, is warm and
dry.

Area and population
Area – 144 000 sq. km
Population – 120 073 000
Population density – 834 per sq. km
Birth rate – 42 per 1000
Death rate – 16 per 1000
Annual rate of population growth – 4.1%
Urban population – 18%
Capital city – Dhaka

Economy

Almost 80% of Bangladesh's population are employed in agriculture. Of the cultivated area, 75% is used to grow rice. The chief cash crop is jute. Bangladesh has large reserves of coal, oil and natural gas, but these are not fully developed. Industry is concerned with the processing of raw materials – jute, limestone, bamboo and softwoods.

GDP – 26 200 million US dollars

GDP per capita – 220 US dollars

BRAZIL

Physical background and climate

A country comprising almost half the area of South America. The north of the country is dominated by the Amazon basin with its tropical rainforests. The land rises to the Guinea highlands in the north and the Brazilian highlands in the south, with large tracts of grassland in between.

Around 93% of Brazil lies between the tropic of Capricorn and the equator, thus the climate ranges from sub-tropical to tropical. Rainfall varies from region to region, with the north east suffering severe droughts.

Area and population
Area – 8 512 000 sq. km
Population – 161 087 000
Population density – 19 per sq. km
Birth rate – 29 per 1000
Death rate – 8 per 1000
Annual rate population growth – 1.4%
Urban population – 78%
Capital city – Brasilia

Economy

Brazil is mainly an agricultural country, the chief crops being sugar cane, manioc, maize, rice, cocoa beans, oranges, coffee (world's largest producer) and soya beans.

Livestock breeding is expanding, much of it taking place on deforested areas of the Amazon rainforest.

Brazil is exceptionally rich in mineral resources: its iron ore deposits are estimated to be the largest in the world. Phosphates, manganese, tin, aluminium, platinum, coal, uranium, gold and copper are also important. 90% of Brazil's power comes from hydroelectric power. A programme for the development of nuclear power stations is well under way.

Brazil was one of the first countries in South America to establish large-scale industry. Iron and steel, car production and machinery are important to the economy.

GDP – 554 600 million US dollars

GDP per capita – 3443 US dollars

CHINA

Physical background and climate

A country in East Asia covering vast areas of land, ranging from low-lying and densely populated plains in the northeast to the high peaks of the Tibetan plateau to the west. It is the second largest country in the world.

More than half the surface area is made up of plateaux and mountains. The Szechuan Basin in central China is highly fertile. China is dissected by three main rivers, the Huang He (Yellow River), the Chang (Yangtse) and Xi (Pearl).

China's climate is dominated by two air masses – polar air brings harsh winters to the north and west. In the summer, warm damp winds bring 80% of the country's rainfall. Altitude obviously has a big effect on temperatures in the mountainous regions.

Area and population
Area – 9 597 000 sq. km
Population – 1 232 083 000
Population density – 128 per sq. km
Birth rate – 18 per 1000
Death rate – 7 per 1000
Annual rate of population growth – 1.1%
Urban population – 38%
Capital city – Beijing

Economy

China is mainly dependent upon agriculture, with an emphasis on food crops: rice and millet in the south, wheat and millet in the north as well as livestock. Cash crops include cotton grown in the north and tea in the south.

Coal is extensively mined and is a major source of power. China is self-sufficient in oil, and small amounts of natural gas are also produced. Iron ore is the most important mineral deposit, and China is the world's leading exporter of tungsten ore. Other minerals include tin, lead, diamonds, gold, manganese and zinc. There has been considerable industrial expansion since 1980, especially in several of the coastal cities. Iron and steel, cotton cloth, locomotives, cement and machinery are important.

GDP – 522 200 million US dollars

GDP per capita – 424 US dollars

EGYPT

Physical background and climate
A country in northeast Africa, extending into
southwest Asia and having borders with Israel,
Libya and Sudan. About 96% of Egypt is desert,
crossed in the east by the River Nile which runs
from south to north, emptying into the
Mediterranean Sea via a huge delta. This deltaic
plain is fringed with lagoons and swamps along
the coast.

Egypt is very hot and dry, and scorching sand
storms can whip through the country in the
springtime. The Mediterranean coastline tends
to be more humid, with milder temperatures.

Area and population
Area – 1 001 000 sq. km
Population – 63 271 000
Population density – 63 per sq. km
Birth rate – 33 per 1000
Death rate – 10 per 1000
Annual rate of population growth – 2.2%
Urban population – 45%
Capital city – Cairo

Economy
Cotton is Egypt's most important cash crop, and
fruit and vegetable production for export has
expanded greatly in recent years. Irrigation is

vital to Egypt's agriculture, with the Nile providing the main source of irrigation water through a network of dams and barrages. Over 50% of the population are farmers.

The Aswan Dam provides electricity to towns throughout the country. Apart from hydroelectric power, Egypt's other main energy source is oil.

Manufacturing is steadily developing, with most industry being located in the delta region – car assembly, a steel mill, and refrigerator and washing machine factories.

Egypt's archaeology makes it an important centre for tourism, which brings in much foreign currency.

GDP – 42 900 million US dollars

GDP per capita – 679 US dollars

FRANCE

Physical background and climate

A country in western Europe bordering Belgium, Luxembourg, Germany, Switzerland, Italy and Spain. It includes the island of Corsica and several overseas regions including Martinique, Guadeloupe and French Guiana.

Fertile lowlands cover most of the north and west of France, rising to the Pyrenees in the southwest, the Massif Central in the south, and the Vosges, Jura and Alps in the east. The coastline is 2700 km in length and the Mediterranean coast has several good harbours. The presence of many lagoons on France's west coast means there are few natural harbours. The main rivers are the Seine, Loire and Rhône.

Climate varies throughout France, with conditions in the Paris Basin being quite different from those along the Mediterranean coast. Winters are generally mild (except in the mountainous areas), and summers are not over-bearingly hot (though high temperatures are common in the south). A particular feature of the French climate is the Mistral wind which blows up the Rhône valley and can do considerable damage to farmland.

Area and population

Area – 547 000 sq. km

Population – 58 333 000
Population density – 106 per sq. km
Birth rate – 14 per 1000
Death rate – 11 per 1000
Annual rate of population growth – 0.3%
Urban population – 74%
Capital city – Paris

Economy
Agriculture is important and the varied landscape means many different crops can be grown and species of livestock kept. Animal products are especially important, as are wheat, sugar, various fruits and vegetables, nuts and butter. The wine industry provides important revenue to the French economy.

The production of natural gas, oil, hydroelectric power and nuclear power have been intensified in recent years, whilst coal production has seen a marked decline. France is a major industrial country – mechanical engineering and industries producing textiles, electrical and electronic goods, vehicles and aircraft, have all prospered since the Second World War.

GDP – 1 540 100 million US dollars

GDP per capita – 26 402 US dollars

GERMANY

Physical background and climate
A country in Central Europe that, until 1990, consisted of the German Democratic Republic (East Germany) and the Federal Republic of Germany (West Germany).

Northern Germany comprises part of the North European plain, while southwards the land rises steadily towards the Black Forest and the foothills of the Alps.

In the coastal regions of Germany the climate is affected by the North and Baltic Seas, giving cold winters and warm summers. In the interior the summers are hotter, with continentality increasing eastwards.

Area and population
Area – 357 000 sq. km
Population – 81 992 000
Population density – 230 per sq. km
Birth rate – data not available
Death rate – data not available
Annual rate of population growth – 0.3%
Urban population – 87%
Capital city – Berlin

Economy

The principal industries are engineering, chemicals, textiles, plastics, motor vehicles and food processing. Agricultural products include grapes (for wine making) hops (for beer making) sugarbeet, barley, wheat, and dariy crops. Western parts of Germany have few natural resources apart from coal and timber while eastern Germany has reserves of iron ore, bauxite, copper, nickel, tin and silver. In addition, banking and international finance are very important. Since unification in 1990, East German industry has been taken over or reorganized by companies from the west, and many jobs have been lost.

GDP – 2 353 200 million US dollars.

GDP per capita – 28 700 US dollars.

GHANA

Physical background and climate

A country in West Africa on the Gulf of Guinea. Nowhere does the land surface rise above 1000 m, and large areas, especially in the centre and north, are large flat plains. The south, however, has a more varied and rugged landscape. Lake Volta in the south of the country is the world's largest artificial lake (7 000 sq. km), formed by the damming of the Volta river at Akosombo.

South west Ghana experiences an equatorial climate, whilst towards the north a seasonal, savanna-type climate prevails.

Area and population

Area – 239 000 sq. km
Population – 17 832 000
Population density – 75 per sq. km
Birth rate – 47 per 1000
Death rate – 15 per 1000
Annual rate of population growth – 3%
Urban population – 36%
Capital city – Accra

Economy

Most of Ghana's population are farmers, with the chief cash crop being cocoa, of which Ghana is the world's third largest producer. Other food crops include maize, yams, cassava and millet. Ghana has many large mineral reserves – gold, manganese, bauxite and diamonds are all exported. However, mining has not provided a basis for industrial development, for although Ghana's largest factory is an aluminium smelter, this depends at present on imported alumina. The smelter is close to the huge dam at Akosombo and uses vast amounts of the hydroelectricity produced there. The loans required by Ghana to develop the Volta scheme have left the country with large debts to the First World.

GDP – 5 400 million US dollars

GDP per capita – 303 US dollars

INDIA

Physical background and climate

A country in South Asia, the seventh largest in the world and the second most populous. India has borders with Pakistan, China, Nepal, Myanmar (Burma) and Bhutan.

The Himalayas, in which the river Ganges rises, form a natural barrier to the north. Central India consists of a large plateau called the Deccan, bordered on either side by the Western and Eastern Ghats. North of this lies the Indo-Gangetic plain with the Thar desert to the west. The range of altitude and latitude in India's landscape have produced a varied climate, though most of India is affected by the monsoon.

Area and population

Area – 3 288 000 sq. km
Population – 994 580 000
Population density – 287 per sq. km
Birth rate – 28 per 1000
Death rate – 11 per 1000
Annual rate of population growth – 1.7%
Urban population – 27%
Capital city – Delhi

Economy

Agriculture employs 70% of India's labour force, with rice, pulses and cereals as the main food crops. Tea, jute, cotton and tobacco are important cash crops. Most agriculture is primitive, with limited technology, knowledge and equipment. Land holdings are often small and uneconomic.

India's large mineral reserves include iron ore, manganese, bauxite and mica. Coal is mined and oil is produced in the Arabian sea. India also has nuclear power.

Manufacturing industries include steel (which is particularly important to the economy), plus textiles, chemicals, fertilizers and engineering. India's chief exports are cotton goods, tea, leather and iron ore.

GDP – 293 600 million US dollars

GDP per capita – 295 US dollars

ITALY

Physical background and climate
A peninsula country in southern Europe having
borders with France, Switzerland, Austria and
Slovenia. The islands of Sicily and Sardinia also
belong to Italy. Most of Italy is rugged and
mountainous with the Apennine range forming a
backbone to the country. There are some flat
coastal areas and the Po valley in the north
forms a wide plain. The climate is generally hot
though there are variations: the mountainous
areas and the Po valley experience harsh
winters, whilst winters in many of the coastal
areas tend to be mild.

Area and population
Area – 301 000 sq. km
Population – 57 266 000
Population density – 190 per sq. km
Birth rate – 12 per 1000
Death rate – 11 per 1000
Annual rate of population increase – 0.1%
Urban population – 68%
Capital city – Rome

Economy
Agriculture is very important, the main crops
being wheat, maize, grapes and olives. 68% of
the land is used for farming. Italy is the world's

largest producer of wine, grapes and pears.

Industry has expanded considerably since the Second World War and is especially concentrated in the wealthier regions to the north of the country. Major industries include machinery and machine tools, steel, motor vehicles and chemicals. Mineral reserves are small. Hydroelectric power is well developed in the Alps and accounts for 33% of electricity generation. Tourism is a very important source of revenue.

GDP – 1 207 000 million US dollars

GDP per capita – 21 077 US dollars

JAPAN

Physical background and climate
A country of eastern Asia consisting of a series of
islands lying between the Pacific Ocean and the
Sea of Japan. The four main islands are Honshu,
Kyushu, Hokkaido and Shikoku. 85% of Japan is
mountainous, with many volcanic peaks. There
is an acute shortage of level land in Japan, with
only 16% of the land available for cultivation.
Most of Japan's lowlands are along the coast.

   The climate of Japan is monsoonal. The winter
monsoon brings rain and snow to the west whilst
the east remains dry but windy. The summer
monsoon brings warm weather throughout the
country.

Area and population
Area – 372 000 sq. km
Population – 125 351 000
Population density – 332 per sq. km
Birth rate – 12 per 1000
Death rate – 7 per 1000
Annual rate of population growth – 0.2%
Urban population – 77%
Capital city – Tokyo

Economy
Rice is the single most important crop – Japan
has to import most of its foodstuffs. Fruits and

vegetables are second in importance after rice.

Although Japan possesses an unusually wide range of minerals, no single mineral exists in large quantities with the exception of sulphur and limestone. Japan relies heavily on imported oil, but is developing a large nuclear power programme. The success of Japan's postwar economic growth has been based upon many factors: a high level of government investment in industry, the Japanese system of labour relations, and a highly efficient education system. Japan is now one of the top three industrial nations in the world. Shipbuilding, cars and hi-tech goods are of paramount importance to the economy.

GDP – 4 599 700 million US dollars

GDP per capita – 36 694 US dollars

KENYA

Physical background and climate
A country in East Africa astride the equator and
bordering the Indian Ocean. It has borders
with Somalia, Ethiopia, Sudan, Uganda and
Tanzania.

The physical environment of Kenya is
extremely diverse. The northern part (almost
half the country) is semi-arid, whilst the narrow
coastal belt is fringed with rainforests and
mangrove swamps. The rest of the country
consists of plateau land covered with vast
expanses of grassland. This upland zone is
bisected by the Great Rift Valley which forms a
north-south trench containing many lakes.
There are two rainy seasons, the 'short rains'
from October to December and the 'long rains'
from March to May. Temperatures are modified
by altitude. Whilst droughts are not uncommon
in many parts of the country, the coastal region
remains humid throughout the year.

Area and population
Area – 583 000 sq. km
Population – 27 799 000
Population density – 48 per sq. km
Birth rate – 54 per 1000
Death rate – 14 per 1000
Annual rate of population growth – 3.4%

Urban population – 28%
Capital city – Nairobi

Economy
Two-thirds of Kenya's population are farmers, mostly at subsistence level. Tea and coffee are important cash crops as are horticultural crops flown direct to markets in Europe.

Mining is still a relatively poor sector, the most important deposit being soda ash. Silver and lead mining is of some importance.

Manufacturing is mainly confined to food processing, though there is a large oil refinery at Mombasa.

In recent years the tourist industry has seen a real 'boom' in Kenya with thousands of tourists visiting the game reserves every year.

GDP – 6 930 million US dollars

GDP per capita – 250 US dollars

MEXICO

Physical background and climate
A country in Central America between the Gulf of Mexico and the Pacific Ocean, and having borders with the United States, Belize and Guatemala. Mountains are the dominating feature of nearly all of Mexico: in the east the Sierra Madre Oriental, and in the west and south the Sierra Madre Occidental. There are coastal plains between the mountain ranges and seas. The climate varies with altitude from the hot coastal lowlands to highland areas where it is too cold for vegetation to grow.

Area and population
Area – 1 973 000 sq. km
Population – 92 718 000
Population density – 47 per sq. km
Birth rate – 31 per 1000
Death rate – 7 per 1000
Annual rate of population increase – 1.8%
Urban population – 75%
Capital city – Mexico City

Economy
Agriculture is still fairly underdeveloped. Maize is the main food crop – cash crops include cotton, sugar, coffee, fruit, vegetables and sisal.

  Fishing is important, and tourism is an

important source of foreign currency.

Mexico now ranks among the world's main oil producing countries. It also has large reserves of natural gas and there are substantial deposits of uranium.

Industrial production has expanded during the last three decades, with steel, car production, electronics and textiles being particularly important.

GDP – 334 200 million US dollars

GDP per capita – 3605 US dollars

NIGERIA

Physical background and climate
A country in West Africa on the Gulf of Guinea, having borders with Niger, Chad, Benin and Cameroon.

Nigeria's main natural feature is the River Niger, Africa's third longest river, which runs north to south, entering the Gulf of Guinea through a broad delta. Behind the lagoons and swamps of the coastal belt is an area of thick forest. The northern half of the country is a plateau giving way to sandy plains merging with the Sahara.

The climatic characteristics of Nigeria summarize much of West Africa – semi-arid in the north and equatorial in the south east.

Area and population
Area – 924 000 sq. km
Population – 115 020 000
Population density – 125 per sq. km
Birth rate – 51 per 1000
Death rate – 16 per 1000
Annual rate of population growth – 2.9%
Urban population – 39%
Capital city – Abuja

Economy
Two-thirds of Nigeria's population work in

agriculture, most of whom are self-sufficient in basic foodstuffs. Chief cash crops are cocoa, groundnuts, rubber and cotton.

Oil production is important around the Niger delta and several small hydroelectric power stations provide energy along the Niger.

Nigeria's industry is expanding steadily with food processing being particularly important.

Textiles and leather are important in the north, and craft industries exist in most towns.

GDP – 35 200 million US dollars

GDP per capita – 306 US dollars

PAKISTAN

Physical background and climate
A country in South Asia bordering on Afghanistan, Iran, China and India.

Pakistan divides into three natural regions: a mountainous belt running north and west (the most famous mountain pass is the Khyber Pass connecting Pakistan with Afghanistan); the high desert plateau of Baluchistan; and a vast alluvial plain drained by the Indus and Punjab rivers, stretching from the Himalayan foothills to the Arabian Sea.

Although classified as having a tropical monsoonal climate, continental characteristics exist throughout the country. Winter is dry and cold in the mountains, but temperatures increase southwards. Summers begin with a hot dry season. Light monsoon rain falls from June to September.

Area and population
Area – 804 000 sq. km
Population – 139 973 000
Population density – 156 per sq. km
Birth rate – 40 per 1000
Death rate – 14 per 1000
Annual rate of population growth – 2.8%
Urban population – 35%
Capital city – Islamabad

Economy

Agriculture is basic to Pakistan's economy and employs around three quarters of the population. Farming is particularly important on the Indus plain where cereals, rice, maize and cotton are grown. Cotton (and cotton cloth) is one of Pakistan's most important exports. Elsewhere, subsistence agriculture persists.

The discovery of natural gas in Baluchistan has played an important role in Pakistan's industrial development.

Manufacturing is becoming increasingly important – chemicals, fertilizers, cement and steel in particular. Pakistan also has many important small-scale industries producing handicrafts.

GDP – 52 000 million US dollars

GDP per capita – 371 US dollars

PERU

Physical background and climate
A country situated in the northwest of South America, bordered by Ecuador, Colombia, Brazil, Bolivia and Chile. In the west, Peru faces the Pacific Ocean and has a 2000 km coastline. It is the fourth largest country in South America.

Three natural zones can be identified – a coastal zone consisting of a narrow strip of desert broken by streams flowing from the Andes; the Andean mountain range (over 6000 m high in places); and the Amazon lowlands to the east.

Coastal Peru is cool and arid, whilst the Andean temperatures vary with altitude, but average around 10°C. The Amazon lowlands are hot and humid all year.

Area and population
Area – 1 285 000 sq. km
Population – 23 944 000
Population density – 19 per sq. km
Birth rate – 34 per 1000
Death rate – 9 per 1000
Annual rate of population growth – 1.8%
Urban population – 72%
Capital city – Lima

Economy
Although 50% of the population is dependent on

agriculture, only 3% of the land is cultivated, and thus Peru is a major importer of agricultural produce. The most productive region is the irrigated coastal belt which produces sugar cane, cotton, rice, vines, bananas, tobacco and citrus fruits.

Minerals are important to Peru's economy, with mines in the Andean region producing silver, copper, lead and zinc.

The manufacturing sector is expanding steadily but most factories are small. The steel and car industries are of particular importance. Peru is a major exporter of fish meal and has a major fishing industry.

GDP – 50 100 million US dollars

GDP per capita – 2092 US dollars

SPAIN

Physical background and climate
A country in southwest Europe occupying more
than four-fifths of the Iberian peninsula. The
Balearic and Canary islands are also part of
Spain. Spain is bordered by Portugal to the west,
and the Pyrenees provide a natural boundary
with France.

Spain's dominant natural feature is the central
Meseta, a vast plateau over 600 m high, crossed
by a series of east-to-west gorges. The Meseta is
bounded by mountains to the north, east and
south.

The northwest coast is mild and wet, whilst the
Mediterranean coast is warmer and drier. The
interior has a continental climate.

Area and population
Area – 505 000 sq. km
Population – 39 674 000
Population density – 79 per sq. km
Birth rate – 15 per 1000
Death rate – 9 per 1000
Annual rate of population growth – 0.1%
Urban population – 76%
Capital city – Madrid

Economy
Except in the humid northern coastal region,

natural conditions for agriculture are poor due to aridity, thin soils and mountainous landscape. The Meseta is the main cereal-producing area, whilst the Mediterranean coast is important for citrus fruits, especially oranges. Spain is the world's leading producer of olives and olive oil. Spain is rich in minerals and has played an historic role as an ore producer in Europe. Nevertheless, Spain cannot supply all its own raw materials and energy requirements. Oil is high on Spain's list of imports.

Important industries include vehicles, ship building (especially important), metals, chemicals and textiles.

GDP – 581 600 million US dollars

GDP per capita – 14 660 US dollars

UNITED KINGDOM

Physical background and climate

The United Kingdom consists of the island of Great Britain, with the Orkney and Shetland Islands and the Hebrides, plus the province of Northern Ireland. It does not include the Channel Islands or the Isle of Man. Great Britain (England, Scotland and Wales) can be divided into lowland and upland areas by an imaginary line running from the mouth of the River Exe in the southwest to the mouth of the River Tees in the northeast. Northern Ireland comprises a broad limestone plain and several small hilly areas, especially along the coast.

The United Kingdom has a temperate, maritime climate.

Area and population
Area – 245 000 sq. km
Population – 58 144 000
Population density – 238 per sq. km
Birth rate – 13 per 1000
Death rate – 12 per 1000
Annual rate of population growth – 0.2%
Urban population – 89%
Capital city – London

Economy
Nearly 80% of the United Kingdom's land is used

for agriculture. Arable farming predominates in the east and southeast, dairying in the southwest, and sheep farming on the uplands.

The main primary sources of energy used by the UK are coal, oil and natural gas, of which the country has many indigenous sources.

Britain is highly industrialized, with over 80% of exports comprising manufactured goods. Metallurgy and engineering employ just under 50% of the industrial workforce; other important industries include the manufacture of vehicles, textiles, aircraft, food processing and alcoholic drinks. In Scotland the electronics industry is important.

GDP – 1 145 800 million US dollars

GDP per capita – 19 706 US dollars

USA

Physical background and climate
A country in North America, the fourth largest in the world. It is a federal republic comprising 50 states, including Alaska in the extreme north west of the continent and Hawaii in the central Pacific ocean.

The Pacific mountain system and the Rocky mountains extend north to south in the western states, with an arid area in between, while the Appalachian mountains extend north to south in the eastern states. In the centre lie vast plains. The USA possesses a wide range of climates, from subtropical in southern Florida to arctic conditions in northern Alaska. Much of the continental interior is dry because of the lack of oceanic influence, though the Great Lakes exert a type of 'oceanic' influence, providing moisture for the surrounding area.

Area and population
Area – 9 373 000 sq. km
Population – 269 444 000
Population density – 29 per sq. km
Birth rate – 16 per 1000
Death rate – 9 per 1000
Annual rate of population growth – 0.9%
Urban population – 76%
Capital city – Washington DC

Economy

Over 50% of the USA is farmland. The vast latitudinal extent of the land mass enables a large variety of crops to be grown and livestock kept. Cereals, cotton and tobacco are the main crops. California and Florida are noted for their fruit growing. Fishing, forestry and associated industries are important.

The USA has huge mineral and fuel reserves, making it a leading producer of natural gas, petroleum, zinc, iron, lead, copper, aluminium, sulphur and electrical energy.

Manufacturing tends to be concentrated in a belt extending from New England to Illinois and Indiana, where complexes of industrial cities are situated, many based around the steel industry, notably cars and electrical engineering, as well as large chemical, textile and food processing industries.

GDP – 7 341 900 million US dollars

GDP per capita – 27 248 US dollars

VENEZUELA

Physical background and climate

A country on the north coast of South America which has a coastline of 3200 km along the Caribbean Sea. It is bordered by Colombia, Brazil and Guyana. It is the sixth largest country in South America.

Venezuela can be divided into four natural regions: the coastal lowlands; the Andes; the Orinoco lowlands; and the Guiana highlands. The coastal lowlands are hot and humid all year round with temperatures over 27°C. Altitude creates a more temperate climate in the Andes region – 16-21°C. 70% of the population live in this region and it contains most of Venezuela's cities, including Caracas. The Orinoco lowlands are a vast treeless plain with sharp seasonal variations in temperature and rainfall. The Guiana highlands in the southeast of the country are sparsely inhabited, containing only 2% of the population.

Area and population
Area – 912 000 sq. km
Population – 22 311 000
Population density – 24 per sq. km
Birth rate – 31 per 1000
Death rate – 5 per 1000
Annual rate of population growth – 2.1%

Urban population – 93%
Capital city – Caracas

Economy
Agriculture is important in the north, with major cash crops including coffee, cocoa, sugar, maize and rice.

However, it is oil which is the mainstay of the Venezuelan economy. Efforts are now being made to diversify the economy as Venezuela is also rich in diamonds, gold, zinc, copper, silver, lead, phosphates and manganese.

Originally manufacturing was centred in Caracas, though several other important centres have now developed.

GDP – 58 300 million US dollars

GDP per capita – 2613 US dollars

APPENDIX 4
National Capitals

Nation	*Capital*
Afghanistan	Kabul
Albania	Tirana
Algeria	Algiers
Andorra	Andorra La Vella
Angola	Luanda
Anguilla	The Valley
Antigua and Barbuda	St. John's
Argentina	Buenos Aires
Armenia	Yerevan
Australia	Canberra
Austria	Vienna
Bahamas	Nassau
Bahrain	Manama
Bangladesh	Dhaka
Barbados	Bridgetown
Belarus	Minsk
Belgium	Brussels
Belize	Belmopan
Benin	Porto Novo
Bermuda	Hamilton
Bhutan	Thimphu
Bolivia	La Paz
Bosnia-Herzegovina	Sarajevo
Botswana	Gaborone
Brazil	Brasilia

Nation	*Capital*
Brunei	Bandar Seri Begawan
Bulgaria	Sofia
Burkina Faso	Ouagadougou
Burundi	Bujumbura
Cambodia	Phnom Penh
Cameroon	Yaoundé
Canada	Ottawa
Cape Verde	Praia
Cayman Islands	Georgetown
Central African Republic	Bangui
Chad	Ndjamena
Chile	Santiago
China	Beijing
Colombia	Bogotá
Comoros	Moroni
Congo	Brazzaville
Congo, Dem. Rep. of	Kinshasa
Costa Rica	San José
Côte d'Ivoire	Yamoussoukro
Croatia	Zagreb
Cuba	Havana
Cyprus	Nicosia
Czech Republic	Prague
Denmark	Copenhagen
Djibouti	Djibouti
Dominica	Roseau
Dominican Republic	Santo Domingo
Ecuador	Quito

Nation	*Capital*
Egypt	Cairo
El Salvador	San Salvador
Equatorial Guinea	Malabo
Eritea	Asmara
Estonia	Tallinn
Ethiopia	Addis Ababa
Falkland Islands	Stanley
Faroe Islands	Tórshavn
Fiji	Suva
Finland	Helsinki
France	Paris
French Guiana	Cayenne
Gabon	Libreville
Gambia	Banjul
Georgia	Tbilisi
Germany	Berlin
Ghana	Accra
Greece	Athens
Greenland	Nuuk
Grenada	St. George's
Guadeloupe	Basse Terre
Guam	Agana
Guatemala	Guatemala City
Guinea	Conakry
Guinea-Bissau	Bissau
Guyana	Georgetown
Haiti	Port-au-Prince
Honduras	Tegucigalpa
Hungary	Budapest

Nation	*Capital*
Iceland	Reykjavik
India	New Delhi
Indonesia	Jakarta
Iran	Tehran
Iraq	Baghdad
Ireland	Dublin
Israel	Jerusalem
Italy	Rome
Jamaica	Kingston
Japan	Tokyo
Jordan	Amman
Kazakstan	Akmola
Kenya	Nairobi
Kiribati	Bairiki
Korea (North)	Pyongyang
Korea (South)	Seoul
Kuwait	Kuwait City
Kyrgyzstan	Bishkek
Laos	Vientiane
Latvia	Riga
Lebanon	Beirut
Lesotho	Maseru
Liberia	Monrovia
Libya	Tripoli
Liechtenstein	Vaduz
Lithuania	Vilnius
Luxembourg	Luxembourg
Macau	Macau
Macedonia	Skopje

Nation	*Capital*
Madagascar	Antananarivo
Malawi	Lilongwe
Malaysia	Kuala Lumpur
Maldive Islands	Malé
Mali	Bamako
Malta	Valletta
Martinique	Fort-de-France
Mauritania	Nouakchott
Mauritius	Port Louis
Mayotte	Dzaoudzi
Mexico	Mexico City
Moldova	Chisinau
Monaco	Monaco-Ville
Mongolia	Ulaanbaator
Montserrat	Plymouth
Morocco	Rabat
Mozambique	Maputo
Myanmar	Rangoon
Namibia	Windhoek
Nepal	Kathmandu
Netherlands	Amsterdam
New Caledonia	Nouméa
New Zealand	Wellington
Nicaragua	Managua
Niger	Niamey
Nigeria	Abuja
Norway	Oslo
Oman	Muscat
Pakistan	Islamabad

Nation	*Capital*
Panama	Panama City
Papua New Guinea	Port Moresby
Paraguay	Asunción
Peru	Lima
Philippines	Manila
Poland	Warsaw
Portugal	Lisbon
Puerto Rico	San Juan
Qatar	Doha
Reunion	St. Denis
Romania	Bucharest
Russia	Moscow
Rwanda	Kigali
Saudi Arabia	Riyadh
Senegal	Dakar
Seychelles	Victoria
Sierra Leone	Freetown
Singapore	Singapore
Slovakia	Bratislava
Slovenia	Ljubjiana
Solomon Islands	Honiara
Somalia	Mogadishu
South Africa	Cape Town/Pretoria
Spain	Madrid
Sri Lanka	Colombo
Sudan	Khartoum
Suriname	Paramaribo
Swaziland	Mbabane
Sweden	Stockholm

Nation	*Capital*
Switzerland	Berne
Syria	Damascus
Taiwan	Taipei
Tajikistan	Dunshambe
Tanzania	Dodoma
Thailand	Bangkok
Togo	Lomé
Tonga	Nuku'alofa
Trinidad and Tobago	Port of Spain
Tunisia	Tunis
Turkey	Ankara
Turkmenistan	Ashkhabad
Uganda	Kampala
Ukraine	Kiev
United Arab Emirates	Abu Dhabi
United Kingdom	London
USA	Washington DC
Uruguay	Montevideo
Vanuatu	Vila
Venezuela	Caracas
Vietnam	Hanoi
Yemen	Sana
Yugoslavia	Belgrade
Zambia	Lusaka
Zimbabwe	Harare

APPENDIX 5
The Geological Time Scale (Approximate)

Million years ago (not consistent scale)	Period	Era	Life forms
	Quaternary: Recent Pleistocene	CENOZOIC	MAMMALS
1	Tertiary: Pliocene Miocene Oligocene Eocene Palaeocene	CENOZOIC	MAMMALS
63			
135	Cretaceous	MESOZOIC	AMPHIBIANS BIRDS REPTILES
181	Jurassic	MESOZOIC	AMPHIBIANS BIRDS REPTILES
230	Triassic	MESOZOIC	AMPHIBIANS BIRDS REPTILES
280	Permian	PALAEOZOIC	FISHES
345	Carboniferous	PALAEOZOIC	FISHES
405	Devonian	PALAEOZOIC	FISHES
425	Silurian	PALAEOZOIC	FISHES
500	Ordovician	PALAEOZOIC	FISHES
600	Cambrian	PALAEOZOIC	FISHES
>600	Pre-Cambrian		

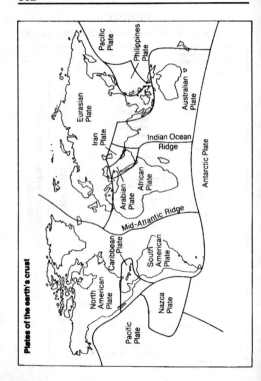

Plates of the earth's crust

Major ocean currents

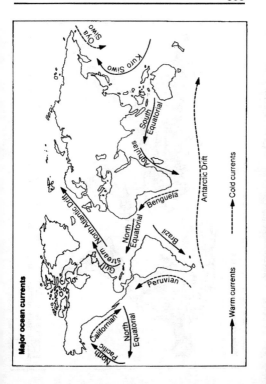

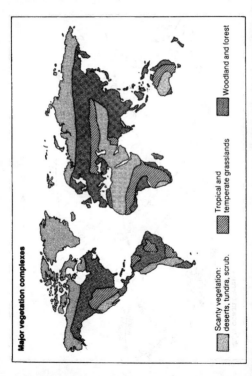

Major vegetation complexes

Scanty vegetation: deserts, tundra, scrub.

Tropical and temperate grasslands

Woodland and forest

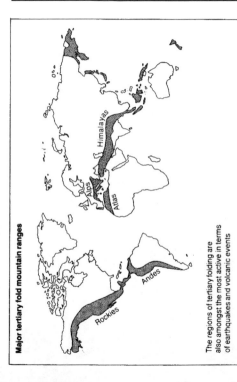

Major tertiary fold mountain ranges

Himalayas

Alps

Atlas

Rockies

Andes

The regions of tertiary folding are also amongst the most active in terms of earthquakes and volcanic events